Production Technology of Vegetables

蔬菜生产技术

汤伟华　刘叶琼◎主编

中国财富出版社有限公司

图书在版编目（CIP）数据

蔬菜生产技术 = Production Technology of Vegetables : 英文 / 汤伟华，刘叶琼主编. —北京：中国财富出版社有限公司，2023.12

ISBN 978-7-5047-8053-9

Ⅰ.①蔬… Ⅱ.①汤… ②刘… Ⅲ.①蔬菜园艺—教材—英文 Ⅳ.① S63

中国国家版本馆 CIP 数据核字（2023）第 238086 号

策划编辑 刘静雯 **责任编辑** 刘静雯 **版权编辑** 李 洋
责任印制 荀 宁 **责任校对** 张营营 **责任发行** 敬 东

出版发行 中国财富出版社有限公司
社 址 北京市丰台区南四环西路188号5区20楼 **邮政编码** 100070
电 话 010-52227588 转 2098（发行部） 010-52227588 转 321（总编室）
010-52227566（24小时读者服务） 010-52227588 转 305（质检部）
网 址 http://www.cfpress.com.cn **排 版** 宝蕾元
经 销 新华书店 **印 刷** 北京九州迅驰传媒文化有限公司
书 号 ISBN 978-7-5047-8053-9/S·0063
开 本 710 mm × 1000 mm 1/16 **版 次** 2025年1月第 1 版
印 张 22.75 **印 次** 2025年1月第 1 次印刷
字 数 396千字 **定 价** 49.80 元

编写人员名单

主　　编：汤伟华　刘叶琼

副 主 编：冯英娜　李小梅

编写人员：汤伟华　江苏农林职业技术学院

刘叶琼　江苏农林职业技术学院

冯英娜　江苏农林职业技术学院

李小梅　黑龙江生物科技职业学院

潘　颀　阿克苏职业技术学院

袁水林　江苏农林职业技术学院

李晓锋　上海农业科学研究院园艺研究所

Foreword

China has abundant vegetable crop resources with a long history of cultivation. In recent years, with the adjustment of vegetable crop structure, the updating of cultivation techniques, great improvements in product quality, and increases in exports volumes and per capita consumption, the demand for horticultural professionals has been rising, especially for those with strong practical skills and higher technical application expertise.

Production Technology of Vegetables is a core course for agricultural technology majors in agricultural and forestry colleges. As a high-quality textbook of the "Double Hundred" quality project at Jiangsu Vocational College of Agriculture and Forestry, it is based on practical vegetable production in China, especially in the southern region. Adhering to the basic principles of higher vocational education, which focus on cultivating the abilities required for frontline positions and job clusters and emphasizing the balance between theoretical teaching and practical training, the course is oriented towards the vegetable production process. The textbook is organized and arranged with typical work tasks in vegetable production as the carrier, taking an integrated approach of teaching, learning, and practical application. There are 8 learning contexts, 23 sub-contexts, and practical training tasks organized and arranged in this textbook. Each typical work task is composed of several implementation stages, aiming to enhance students' abilities in analyzing and solving problems throughout the learning process. This textbook is applicable to students majoring in horticultural technology.

This textbook is co-edited by Associate Professor Tang Weihua and Lecturer Liu Yeqiong from Jiangsu Vocational College of Agriculture and Forestry. The contributors and their respective responsibilities are as follows: Tang Weihua is responsible for the production of melon vegetables and solanaceous vegetables; Yuan Shuilin from Jiangsu Vocational College of Agriculture and Forestry is responsible for the production of cabbage vegetables; Pan Qi from Aksu Vocational and Technical College is responsible for the production of root vegetables; Feng Yingna from Jiangsu Vocational College of Agriculture and Forestry is responsible for the production of leafy vegetables; Liu Yeqiong is responsible for the production of allium vegetables; Li Xiaomei from Heilongjiang Agricultural Engineering Vocational College is responsible for the production of tuber vegetables; Li Xiaofeng from Shanghai Academy of Agricultural Sciences is responsible for the production of legume vegetables. After the manuscript was completed, it was finalized by the editors-in-chief and reviewed by Professor Yan Zhiming from Jiangsu Vocational College of Agriculture and Forestry.

The compilation of this textbook was supported and guided by Jiangsu Vocational College of Agriculture and Forestry and China Fortune Press Co., Ltd. with reference to relevant literature and material. Therefore, we would like to express our gratitude for their contributions.

Due to limitations in knowledge and expertise, as well as time constraints, there may be inevitable omissions and shortcomings in the content. We sincerely invite experts, scholars, and readers to provide valuable feedback and share their experiences to help improve the content.

Editors

January 2025

Contents

Learning Context 1 Production of Melon Vegetables ······ 1

Overview ······ 2

Sub-context 1 Production of Cucumber ······ 5

Sub-context 2 Production of Watermelon ······ 19

Sub-context 3 Production of Netted Muskmelon ······ 30

Learning Context 2 Production of Solanaceous Vegetables ······ 41

Overview ······ 42

Sub-context 1 Production of Tomato ······ 45

Sub-context 2 Production of Eggplant ······ 64

Sub-context 3 Production of Chili Pepper ······ 80

Learning Context 3 Production of Legume Vegetables ······ 95

Overview ······ 96

Sub-context 1 Production of Cowpea ······ 97

Sub-context 2 Production of Kidney Bean ······ 105

Sub-context 3 Production of Vegetable Soybean ······ 113

Learning Context 4 Production of Cabbage Vegetables ······ 121

Overview ······ 122

Sub-context 1 Production of Chinese Cabbage ······ 124

Sub-context 2 Production of Head Cabbage ······ 140

Sub-context 3 Production of Cauliflower ······ 152

Sub-context 4 Production of Chinese Kale ······ 161

Learning Context 5 Production of Root Vegetables ········· 168
Overview ········· 169
Sub-context 1 Production of Radish ········· 171
Sub-context 2 Production of Carrot ········· 183

Learning Context 6 Production of Leafy Vegetables ········· 194
Overview ········· 195
Sub-context 1 Production of Lettuce ········· 198
Sub-context 2 Production of Celery ········· 208

Learning Context 7 Production of Allium Vegetables ········· 223
Overview ········· 224
Sub-context 1 Production of Chinese Chive ········· 225
Sub-context 2 Production of Garlic ········· 238
Sub-context 3 Production of Onion ········· 249

Learning Context 8 Production of Tuberous Vegetables ········· 262
Overview ········· 263
Sub-context 1 Production of Potato ········· 264
Sub-context 2 Production of Ginger ········· 277
Sub-context 3 Production of Taro ········· 287

Practical Training Tasks ········· 299

References ········· 355

Learning Context 1　Production of Melon Vegetables

「**Learning Objectives in This Context**」

Understand the general cultivation characteristics of melon vegetables. Master the biological characteristics, variety types, and cultivation seasons of cucumber, watermelon, netted muskmelon, winter melon, and zucchini. Acquire the techniques for early cultivation of cucumber, watermelon, netted muskmelon, winter melon, and zucchini in the greenhouse.

「**Analysis of Tasks in This Context**」

Master the tasks involved in soaking and germination, sowing and seedling raising, land preparation, mulching film covering, transplanting, plant regulation, fertilization and irrigation management, hand pollination, and fruit thinning, as well as pest and disease control for cucumber, watermelon, netted muskmelon, winter melon, and zucchini.

「**Introduction**」

Melon vegetables hold a significant position in China's horticultural industry. With various types, they are essential ingredients in people's lives. They not only serve as an important source of vegetables for us but also the preferred vegetable types for farmers in many regions to increase their income and agricultural efficiency.

Overview

I. Types of Melon Vegetables

Melon vegetables refer to the cultivated plants of the Cucurbitaceae with edible fruit. There are many varieties cultivated in China, including bottle gourd, pumpkin, winter melon, luffa, muskmelon, watermelon, gourd, bitter gourd, chayote, snake gourd, etc. (Fig. 1-1). They mainly comprise nine genera: *Cucurbita*, *Luffa*, *Benincasa*, *Lagenaria*, *Citrullus*, *Cucumis*, *Sechium*, *Trichosanthes*, and *Momordica*.

Fig. 1-1 Different Melon Vegetables

Chayote Snake gourd Gourd

Fig. 1-1 Different Melon Vegetables (continued)

Most melon vegetables are annual herbaceous plants, with the exception of chayote, which is a perennial one. They share many biological and cultivation commonalities.

II. Biological Commonalities of Melon Vegetables

Melon vegetables are creeping plants. Their stems are hollow and can grow up to several meters, covered with coarse hairs or spines. Their nodes have tendrils, which are modified branches or leaves used for climbing. The leaves are large and alternate, with relatively long petioles. The leaf blades are slightly heart-shaped or palmately lobed. In each leaf axil of the main vine, lateral vines (known as side shoots) can grow, which can then produce additional lateral vines (referred to as secondary shoots). When the vine of the melon vegetable comes into contact with soil, it can easily develop adventitious roots on the nodes.

Melon vegetables are monoecious plants, with some cultivars having bisexual flowers, male plants, and female plants. Generally, there are more staminate flowers that appear earlier than the solitary pistillate flowers. Both staminate and pistillate flowers have nectaries and are insect-pollinated, making them naturally cross-pollinated. The corolla of most flowers is yellow, while gourds have white flowers that open at night. If it is rainy or at low temperatures during the flowering period, insect activity may be limited. Therefore, hand pollination is required to ensure a successful fruit set.

Chayote has one seed per fruit, while other melon vegetables have many seeds.

Melon vegetables prefer warm temperatures and have low cold tolerance. The optimal temperature range for their growth is typically between 20–30°C, with poor growth observed below 15°C and growth cessation occurring below 10°C. They start to suffer damage below 5°C, so cultivation must be carried out during the frost-free period in Zhejiang. Proper cooler temperatures during the seedling stage can promote the formation of pistillate flowers.

The fruiting habits of melon vegetables are generally divided into three categories: in the first category, the fruits are mainly produced from the main vine, such as early-maturing cucumbers and zucchini; in the second category, the fruits are mainly produced on lateral vines, such as muskmelon and pumpkin, which typically utilize lateral vines for fruiting; in the third category, the fruits are produced on both the main vine and lateral vines, such as winter melon, pumpkin, luffa, watermelon, and bitter gourd.

III. The General Cultivation Principles of Melon Vegetables

Melon vegetables, except for cucumbers, have well-developed root systems, but their regenerative ability is weak. After transplanting, seedlings have a long period of acclimation, so they are suitable for direct sowing, transplanting young seedlings, or taking root protection measures for seedling raising.

Techniques such as pruning, pressing vines with soil, or setting up trellises are commonly used for the cultivation of melon vegetables.

Melon vegetable seeds are large and have abundant nutrient reserves. The cotyledons can undergo photosynthesis before the true leaves unfold, so it is important to protect the cotyledons. During germination, a common issue for melon vegetables is that the seed coat does not easily shed, and the cotyledons become trapped inside the seed coat, resulting in deformities and affecting photosynthesis. To facilitate the easy shedding of seed coat, seeds should be laid flat during sowing, and the soil covering them should not be too thin. It is also necessary to maintain a higher temperature and adequate soil humidity when the seedlings emerge.

Melon vegetables belong to the Cucurbitaceae and share many common diseases, such as fusarium wilt, blight, downy mildew, anthracnose, and powdery mildew. Therefore, different types of melon vegetables should not be grown consecutively but rotated with other vegetables or crops.

Sub-context 1 Production of Cucumber

Cucumis sativus, also known as cucumber, can be consumed fresh, cooked, or pickled, and each cooking method has its own style, making it popular among the general public. It is widely cultivated throughout China, both in open fields and protected areas, with rich genetic resources. Cucumber production can be carried out almost year-round, ensuring a consistent supply.

I. Biological Characteristics

1. Botanical features

(1) Roots: Cucumber has a shallow root system, with roots distributed in the top 20 cm plow layer of soil. The roots are shallowly embedded and exhibit a frail and delicate nature, which indicates a weak absorption capability.

(2) Stems: The main vine is 2–3 m long, with relatively weak branching ability. During the seedling stage, the stem is short, compact, and almost upright. After the emergence of 4–5 true leaves, the internodes elongate, adopting a vining nature, with flowers and tendrils emerging from leaf axils (Fig. 1-2).

(3) Leaves: Cotyledons arise opposite to each other, with an alternate arrangement of true leaves. The petioles are long, and the leaf blades are palmatilobate, large, and thin, which are characteristics of strong transpiration. The leaf margins are finely serrated and covered with fine hairs. Leave axils have axillary buds or floral bud primordia, and tendrils appear after vine growth.

(4) Flowers: Flowers are axillary and monoecious. Occasionally, there are staminate flowers and pistillate bisexual flowers on the same plant. Additionally, there are exclusively female plants. Staminate flowers usually blossom earlier than pistillate flowers, with alternating blossoming. Both pistillate and staminate flowers have nectaries and are pollinated by insects. The natural hybridization rate between

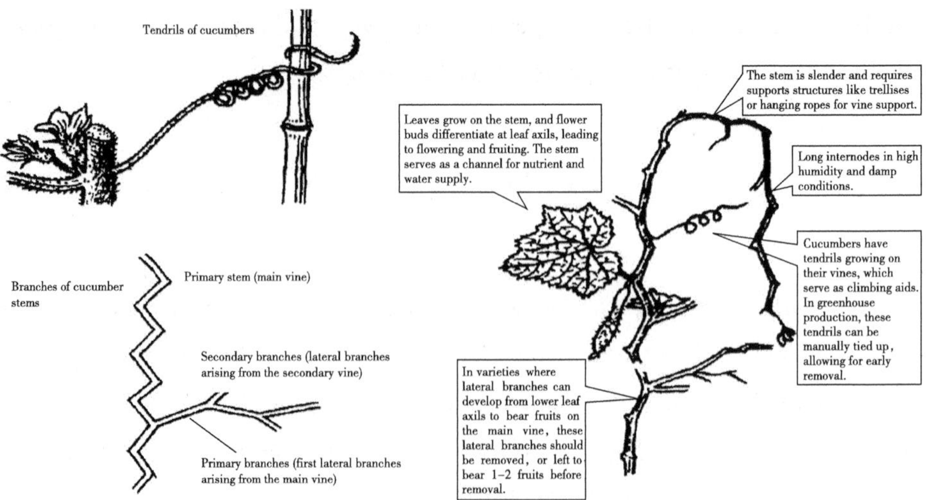

Fig. 1-2 Cucumber Stems and Vines

varieties ranges from 53%-76%.

The main factors that affect sexual differentiation in cucumbers are as follows.

① Temperature and day length: Big temperature differences between day and night and short day length are good for the differentiation of pistillate flowers.

② Different mineral nutrition: Nitrogen promotes the growth of pistillate flowers, while potassium nutrition favors the formation of staminate flowers.

③ Air humidity and carbon dioxide concentration: High air humidity and high carbon dioxide concentration favor the formation of pistillate flowers.

④ Foliar application of ethephon increases the number of pistillate flowers, while gibberellin application significantly increases the number of staminate flowers.

(5) Fruits: The fruits of cucumbers are spurious fruits, developed from the ovary and receptacle. They are green or white, wax-coated, and have smooth surfaces or ridges, bumps, or spines. Fruit shape, color, ridges, bumps, spines, and other characteristics serve as important criteria for varietal identification. Generally, their young fruits can be harvested 7-15 days after flowering. It takes about 35-45 days from flowering to fruit maturity.

(6) Seeds: The seeds are lanceolate, flat, yellowish-white, with thousand-seed

weight of 22–42 g. Germination viability lasts 4–5 years.

2. Requirements for environmental conditions

(1) Temperature: Cucumbers are thermophilic crops with an optimal growth temperature of 25–30°C. They are not cold-tolerant, and the optimal temperature for photosynthesis is 25–32°C. The suitable temperature for seed germination is 27–29°C. During the seedling stage, the daytime temperature should be 22–28°C, while the nighttime temperature should be 17–18°C or above 13°C. During the flowering and fruiting stage, the daytime temperature should be 25–29°C, and the nighttime temperature should be 18–22°C. After the peak harvest period, the temperature should be slightly lower to prevent plant aging and maintain a longer harvest period. The suitable soil temperature for cucumber root system growth is 20–23°C, and cucumbers stop growing at 10–12°C, so the temperature must be kept above 15°C.

(2) Light intensity: Cucumbers require strong light, but they can also adapt to weaker light. Therefore, they are suitable for protected cultivation in winter and spring. However, insufficient light has adverse impacts on its yield and quality. Shorter daylight hours are favorable for the formation of pistillate flowers, but different varieties have different responses to short daylight hours.

(3) Water moisture: Cucumbers have a large leaf area and high transpiration rate, while their root system is shallow, with a small distribution range and weak absorption capacity. Therefore, they require higher soil humidity and air humidity. Cucumbers are not drought-tolerant crops, and they require sufficient soil moisture and humid air.

(4) Gases: The carbon dioxide saturation concentration for cucumber photosynthesis is 0.1%, and under sufficient light, high temperature, and high humidity conditions, the saturation concentration of carbon dioxide can reach around 1%. The air only contains 0.03% carbon dioxide. In cultivation, methods such as applying organic fertilizers, improving ventilation, and supplementing carbon dioxide can be adopted to increase the carbon dioxide content in the air.

(5) Soil: Cucumbers grow rapidly and have high yields, requiring well-structured loamy soil to overcome the characteristics of being moisture-loving but intolerant to waterlogging, and nutrient-loving but not tolerant to excessive fertilization. Among

the absorption of three essential elements, cucumbers absorb potassium the most, followed by nitrogen, and phosphorus the least. The ratio among the three elements is approximately 2.5 ∶ 1 ∶ 4.0. The optimal soil acidity is neutral or slightly acidic (with a pH value of 5.5–7.2) for good growth. The concentration of soil solution should not be too high. It is advisable to choose loamy soil or sandy loam soil that retains water and nutrients well, and is rich in organic matter for cultivation.

II. Types and Varieties

Ecological types of cucumbers are as follows.

(1) Cucumber in South China: The cucumber has lush stem and leaves, and is tolerant to dampness and heat. It is a short-day plant with small, sparse bumps and more black spines. The young fruit is green, greenish-white or yellowish-white in color, with a mild taste. The mature fruit is yellowish-brown with net-like patterns. It is distributed in the South of the Yangtze River basin in China and various parts of Japan. Representative varieties include Kunming Early Spring Burpless Cucumber, Guangzhou Erqing, and Shanghai Yanghang.

(2) Cucumber in North China (Fig. 1-3): The plant has moderate growth vigor and prefers such natural conditions as moist soil and sunny weather. It is not sensitive to the length of sunlight. The young fruit is cylindrical in shape and green in color, densely covered with bumps and numerous white spines. The mature fruit is yellowish-white without net-like patterns. It is distributed in the north of the Yellow River basin in China, as well as in North Korea, Japan, and other regions. Representative varieties include Shandong Xintai Mici, as well as hybrids Zhongnong 1101, Jinza No.1, No.2, etc.

(3) Cucumber in South Asia (Fig. 1-4): The stem and leaves are thick and easily branched. The fruit is big, with a single fruit weighing 1–5 kg, and is short or long cylindrical in shape. The fruit skin is light with sparse bumps, and the spines are black or white. The fruit has thick skin and is tasteless. It prefers a warm and humid climate and strictly requires short daylight hours. They are distributed in various regions of South Asia, with many local varieties, such as Sikkim, Chinese Banna, and Zhaotong.

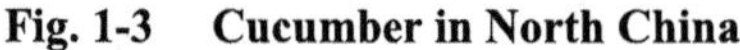

Fig. 1-3 Cucumber in North China

Fig. 1-4 Cucumber in South Asia

(4) Open-field cucumber in Europe and America: The cucumber has lush stem and leaves, and the fruit is cylindrical and medium-sized, with sparse bumps, white spines, and a mild taste. The mature fruit is light yellow or yellowish-brown. There are variety groups of Eastern Europe, Northern Europe, North America, etc., distributed in various regions of Europe and North America. In recent years, some varieties have been introduced and cultivated in China.

(5) Greenhouse cucumber in Northern Europe: The cucumber has lush stem and leaves, being tolerant to low temperatures and weak light. The fruit has a smooth surface, light green in color, and can reach lengths of over 50 cm. It is distributed in the UK and the Netherlands, with representative varieties including English Greenhouse Cucumber, Dutch Greenhouse Cucumber, etc.

(6) Small-sized Cucumber: Their plants are relatively short and highly branched, with abundant flowers and fruits. They are distributed in various regions of Asia, Europe, and America, with representative varieties including Yangzhou Tender Cucumber.

III. Cultivation Season and Methods

In the Yangtze River basin and its southern regions, where the frost-free period is long, cucumbers can be grown throughout the year. Cultivation in open fields is primarily conducted in the summer and autumn, while in early spring, polytunnels and other facilities are often used for protected cultivation (Table 1-1).

Table 1-1 Cultivation Season and Cropping Schedule for Cucumbers in the Yangtze River Basin

Season and Cropping Schedule	Sowing Phase	Transplanting Phase	Harvest Phase
Spring Crop	Late February – Early March	Early April – Late April	Late April – Late June
Summer Crop	Late May – Mid-June	Direct sowing	Early July – Late August
Autumn Crop	Late July – Early August	Direct sowing	Early September – Late October

IV. Cultivation Techniques of Early-maturing Cucumbers in Polytunnels in Spring

1. Variety selection

Early-maturing varieties should be selected for the cultivation of early-maturing cucumbers in polytunnels. The characteristics of early-maturing varieties are as follows: They have main vine fruits and low-lying root fruits, with short internodes, tolerance to certain weak light conditions, adaptation to big temperature differences, and resistance to various diseases. According to local conditions and dietary habits, the main varieties currently selected for production are early-maturing varieties, such as Changchun Mici, Xintai Mici, Jinyou No.2, Jinyou No.3, Jinlv No.3, Jinchun No. 3, Zhongnong 12, etc.

2. Nurture of robust seedlings

Grafting of cucumber seedlings is one of the yield-increasing measures, especially in protected cultivation of cucumbers.

Criteria for robust seedlings: The plant is 15–20 cm high, with 4–5 true leaves, vigorous and healthy, with intact cotyledons. Its leaves should be large and thick, deep green in color. Its stems should be sturdy with short internodes and a well-developed root system. The seedlings should be free from pests and diseases, and should be around 40–50 days old. Due to the long duration of seedling growth and the larger size of the seedlings, root protection measures should be taken to minimize root damage during transplanting and shorten the acclimation period. The seedlings should be hardened off before transplanting.

Temperature management during seedling stage: Before emergence, the soil temperature should be around 25–30°C during the day, and 15–18°C at night. After emergence, the temperature should be maintained at 25–30°C in the daytime and around 13–17°C at night, while the soil temperature should be kept above 15°C. Low-temperature hardening should be carried out for 7–10 days before transplanting of seedlings.

Moisture management: Soil moisture should be maintained at a suitable level. To ensure normal growth of seedlings, it is better to keep the soil slightly dry rather than excessively wet, which can help improve the quality of the seedlings.

3. Preparations before transplanting

To increase soil temperature for early planting, it is recommended to cover the polytunnel with plastic film 20 days before transplanting. The ridge should be 1 m wide with deep furrows that are 30–40 cm in depth. Apply 5,000 kg of decomposed poultry or livestock manure and 25 kg of compound fertilizer per mu (1 mu≈ 667 m^2), either by opening furrows at the center of the ridge or by full-layer application. Cover the polytunnel with plastic film before transplanting, with a mulching film width of 140–150 cm, covering both the ridge back and furrow completely.

4. Transplanting

(1) Transplanting period: When the air temperature inside the polytunnel stabilizes above 10°C with the minimum temperature above 0°C, and a soil temperature at a depth of 10 cm is stable between 10–12°C, it is considered a safe planting period. For polytunnel transplanting in early spring, it is crucial to understand the weather patterns and changes. During the safe transplanting period, we should take advantage of the period at the end of cold front and the beginning

of warm front based on the characteristics of “three colds and four warms” in early spring. Transplant the seedlings at the beginning of the warming period, striving to have the seedlings well-acclimated before the next cold wave arrives.

(2) Transplanting density: When the film is uncovered for transplanting, it is recommended to transplant the cucumber plants at around 4,000 plants per mu, with spacing at 60 cm × 30 cm, or by enlarging row spacing and narrowing plant spacing, i.e. (100–120) cm × 18 cm, which is beneficial for relay intercropping. Another option is cultivating primary and secondary rows, with a ridge width of 100–120 cm and double rows per ridge. The primary row has a plant spacing of 18 cm, while the secondary row has a plant spacing of 30–35 cm. When the cucumbers in the secondary row have 12–14 leaves, they should be topped, and after yielding 2–4 fruits, the seedlings should be removed, for the purposes of increasing early yield in the secondary row and cultivating the cucumbers in the primary row. Alternatively, the beds can be raised with bed spacing set at 50 cm and 80 cm. The small furrow of the ridge is approximately 20 cm wide and 10–15 cm deep, mainly for watering and topdressing in cool seasons. The big furrow of the ridge is about 35–40 cm wide and 20 cm deep, mainly for field access and watering during the high-temperature periods.

(3) Measures for early transplanting of greenhouse cucumbers: For cucumber greenhouse cultivation in early spring, the key to improving economic benefits lies in early transplanting. However, due to the cold weather in early spring, early transplanting requires well-established cold-resistant and insulation facilities; otherwise you may suffer from significant economic losses. Ideal cold-resistant and insulation facilities include multiple layers of covering, such as covering a smaller greenhouse with a bigger polytunnel, creating a smaller or micro greenhouse inside a bigger greenhouse, adding non-woven fabric or plastic film as an outer layer in addition to the double-layer shade cloth and greenhouse film at nighttime, or temporarily heating the greenhouse with an air heater or stove before the arrival of cold waves. All these methods can increase the temperature inside the greenhouse by 2.4–3.6°C and prevent damage from cold waves.

5. Field management

(1) Fertilizer and water management: After the base fertilizer is applied to the cucumber, no topdressing should be applied before fruiting. However, the seedling-boosting fertilizer can be applied once. After fruiting begins, fertilization should be combined with watering. Generally, fertilization should be applied every 5 days, alternating between chemical fertilizers (compound fertilizers) and organic fertilizers (cake fertilizer, chicken manure, human manure urine composting solution). During the peak period of fruiting, extraradical topdressing can be carried out in combination with sprayings, such as 0.2% urea, 0.1% potassium dihydrogen phosphate, 0.1%–0.2% magnesium sulfate, 0.2% boric acid solution, etc.

(2) Plant regulation: In addition to normal trellising, vine tying, branch pruning, and tendril removal, there are several new measures for early maturing greenhouse cultivation. ① Vine direction: In order to reduce shading, save trellis materials, and make full use of space, the method of tying the vine to bamboo trellising can be replaced by the method of hanging the vine with white plastic strings, that is, tying the two ends of iron wires to the frame of polytunnel along the planting row, connect the strings to the cucumber roots and stems at the hole spacing from top to bottom. The tendril can be directed by turning it in one direction and wrapping it with a plastic string. ② Pinching: The main vine can be pinched when it is about to touch the greenhouse film to not only prevent the stem tip from getting scorched after touching the film, but also eliminate the apical dominance and promote the growth of cucumber on the side shoots. Lateral vines can also be pinched after one fruit and one leaf is kept to significantly increase the yield. ③ Leaf picking: Removing older yellow leaves and diseased leaves from the lower part of the plant in a timely manner can reduce nutrient consumption and improve ventilation and light transmission conditions (Fig. 1-5). ④ Lowering vines: When the vine reaches the top of the rope, the older leaves and cucumbers at the base of the vine should be removed before the vine is lowered. Then, loosen the string at the top of the vine and gently guide it downwards, allowing it to coil around the mulching film. Ensure the vine is lowered to a height where the functional leaves remain above the ground. Tie the top string to hold the cucumber vine, and

periodically lower the vines (Fig. 1-6). During the low-temperature period, it is recommended that cucumber plants be regulated on sunny and warm days between 10:00 and 15:00. This can take advantage of the higher temperatures in this time window to facilitate wound healing and prevent disease. For the high-temperature period, vine wrapping should be conducted in the afternoon when the cucumber vines have softened due to water loss, for wrapping the vines in the morning can potentially cause damage to the stems and leaves.

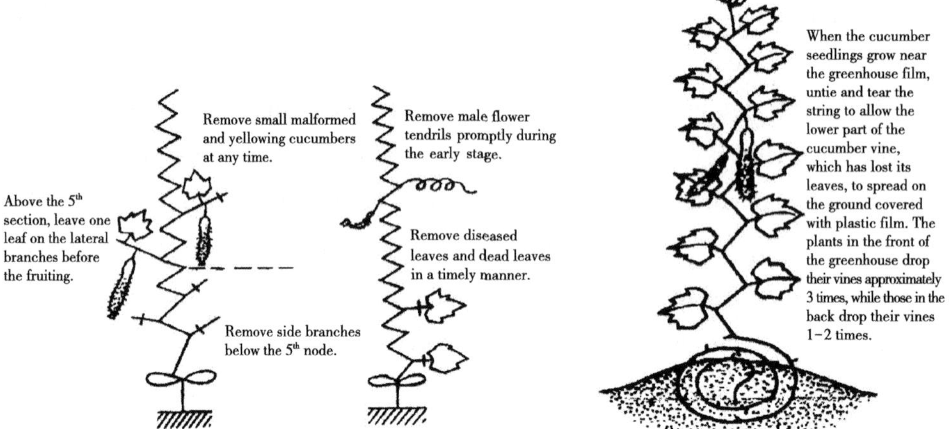

Fig. 1-5 Pruning of Cucumber **Fig. 1-6 Lowering Vines of Cucumber**

(3) Application of plant growth regulators: Spraying 100–200 μL/L ethephon when cucumber seedlings have developed the first to third true leaves can significantly increase the number of pistillate flowers. After the pistillate flowers are increased, the amount of fertilization must be increased; Otherwise, a large number of cucumbers may experience fruit abortion. According to international reports, treating the ovaries of pistillate flowers at different stages of development with plant growth regulators, such as gibberellin, 2,4-D, etc., has been found to promote fruit development.

6. Pest and disease control

Cucumbers are prone to various diseases, including fusarium wilt, blight, downy mildew, anthracnose, and powdery mildew, of which, the first four diseases, can have a severe impact on cucumber production. Bacterial angular leaf spot disease and viruses also pose significant threats to cucumber plants, greatly affecting their yield. Therefore, in addition to timely application of fungicides for disease prevention and control, attention should be paid to the selection of disease-resistant varieties, crop rotation, cultivating strong seedlings with nutrient bowls or nutrient blocks, avoiding root damage during transplantation, and adopting high ridge cultivation and good drainage to promote the development of root systems, enhance the plant growth and increase the disease resistance, which can greatly reduce or prevent the occurrence and spread of diseases and pests.

Grafting with larger seedlings is commonly practiced in cucumber cultivation, with pumpkin or bottle gourd serving as the rootstock. Grafting can increase the plant's disease resistance and stress tolerance, and promote the healthy development of the root system, which has a positive effect on preventing root-caused diseases.

Common pests include cucumber beetles, aphids, greenhouse whiteflies, and *Liriomyza sativae*. Early intervention is essential to prevent leaves from damage and virus infection.

7. Timely harvest

Cucumbers must be harvested when they are young, particularly for root cucumbers, which should be harvested 8–10 days after flowering. The criteria for harvesting cucumbers are the presence of flowers on the top of the cucumber, and tender thorns and prominent stripes on the main body. After harvest, they should be sorted and packed in separate grades to prevent friction and ensure the quality of the produce.

V. Common Cultivation Issues and Their Prevention Strategies

1. Early fruit abortion of spring cucumbers

The pistillate flowers of the cucumbers stop growing and gradually turn yellow and wither, which is called fruit abortion. Primary reasons: Varieties with poor parthenocarpy, excessive planting density, excessively low or high temperatures, continuous rainy days with small temperature differences between day and night, inadequate water and fertilizer supply, etc. Preventive measures: Firstly, select excellent varieties with good parthenocarpy, such as Changchun Mici and Xintai

Mici. Secondly, maintain daytime temperatures at 20–35°C and nighttime temperatures at 10–20°C. Thirdly, enhance water and fertilizer management, promptly prune direct vines, and improve ventilation and light transmission conditions.

2. Misshapen cucumbers

① Curly cucumber: The curly cucumber has a crescent moon shape. During the seedling cultivation process, particularly during the stage of flower bud differentiation to flowering, the inadequate supply of temperature, water, and nutrients can lead to the formation of curved ovaries, which eventually develop into curly cucumbers. Insufficient fertilizer supply and drought during the fruit enlargement period can also lead to the formation of curly cucumbers. Therefore, it is important to create favorable conditions for flower bud differentiation during the seedling stage. During the fruit enlargement stage, enhanced proper management with sufficient water and fertilizer can prevent the occurrence of curly cucumbers.

② Big-belly cucumber: Due to incomplete fertilization in cucumbers, only the central to the tip of the fruit is fertilized. The fertilized part grows faster, while the unfertilized part grows slowly, resulting in the formation of big-belly cucumber. Additionally, the relaxation of field management in the later stage, causing weak growth of plants, can also lead to the formation of big-belly cucumbers. Lack of potassium in the soil is another cause of big-belly cucumber. Therefore, for the cucumber varieties with poor parthenocarpy that require pollination, it is necessary to prevent incomplete fertilization by hand pollination and beekeeping in greenhouses. During the period from peak fruiting to its later stage, water and fertilizer management should be strengthened. Additionally, topdressing of 0.2% potassium dihydrogen phosphate can be combined to prevent the occurrence of big-belly cucumber.

③ Pointy cucumber: It is caused by poor fertilization. The continuous high temperature and drought in the greenhouse are important factors leading to the production of pointy cucumbers. Improper water and fertilizer management can also lead to the formation of pointy cucumbers due to weak plant growth. Therefore, if the temperature exceeds 30°C in protected cultivation, immediate ventilation should be provided to prevent excessive heat. During the fruiting period, the soil should be kept moist, and water and fertilizer management should be enhanced. As long as the plants are vigorous and healthy, pointy cucumber can be prevented.

④ Narrow-waisted cucumber: It is caused by long-term high temperature and dryness in the greenhouse, weak plant growth vigor, and potassium deficiency in the soil. The narrow-waisted cucumber can be prevented when you pay more attention to water and fertilizer management.

⑤ Thin-shoulder cucumber: Thin-shoulder cucumber refers to the cucumber with a short stem and slender and elongated shoulder. This is caused by prolonged low temperatures at nighttime or excessive nutrition. As long as the temperature in the greenhouse is maintained above 10°C and water and fertilizer management is enhanced, the cucumber may quickly return to normal.

3. Bitter taste in cucumber fruits

Cucumber fruits may sometimes have a bitter taste, which is caused by bitter substances called cucurbitacin. Such bitterness in fruits generally occurs near the stem and shoulder, and it is related to genetics, as well as physiological and environmental conditions. Fruits grown on older plants and overripe fruits may have a bitter taste, and those produced from plants with damaged root system or poorly growing plants also tend to have a bitter taste. Long-term low temperatures, insufficient sunlight, high temperature and drought, excessive nitrogen fertilizer application, and nutrient deficiency can also lead to bitter fruit. During drought periods, fruits produced under low temperatures can also have a bitter taste. To overcome the bitterness, plants with bitter fruits should be eliminated, and cultivation management should also be improved to promote normal plant growth and fruit development.

4. Cucumber flower topping-off

Flower topping-off is mainly manifested as the stunted growth of the growing point and several shortened nodes of the tip, forming a bump, and the staminate and pistillate flowers continue to bloom until the tip of the cucumber seedling. In production, it may be caused by low nighttime temperatures, big temperature differences between day and night, and excessive formation of pistillate flowers, which inhibits the vegetative growth. Low ground temperature, excessively dry or wet soil, and excessive fertilization can cause poor root system development and weak nutrient absorption in cucumbers. In order to prevent flower toppling, we can take such measures as increasing leaf temperature, strengthening water and fertilizer management, timely intertilling and loosening the soil, and promoting the development of root system. For plants that have already experienced flower topping-off, timely harvest of marketable fruits and removal of some pistillate flowers are necessary. Generally, leave 1–2 fruits per plant on strong plants, and remove all fruits on weak plants.

「**Summary of Sub-context**」

As cucumbers are vegetables with relatively weak nutrient absorption ability, it is advocated to apply less fertilizer but more frequently, and timely harvest with sufficient water supply.

「**Expanded Knowledge**」

Therapeutic Effects of Cucumbers

(1) Improve body immunity. The cucurbitacin C in cucumbers can improve our body's immune function.

(2) Anti-aging. Cucumbers are rich in vitamin E and cucumber enzyme, which can promote longevity and have anti-aging effects.

(3) Prevention of alcohol poisoning. The alanine, arginine, and glutamine contained in cucumbers have certain auxiliary treatment effects on patients with alcoholic hepatitis, and can prevent alcohol poisoning.

(4) Lowering of blood sugar. Diabetics can take cucumbers instead of starchy foods to satisfy hunger, as cucumbers, far from raising blood sugar levels, can even lower them.

(5) Promote weight loss. The tartronic acid, contained in cucumbers, can inhibit the conversion of carbohydrates into lipids.

(6) Brain nourishing and calming. Cucumbers contain vitamin B1, which is beneficial for improving brain and nervous system function and can assist in the treatment of insomnia.

Sub-context 2 Production of Watermelon

Watermelon originated in the tropical grasslands of Africa. The flesh is crisp, tender, juicy and nutritious. It is an important summer fruit and has been cultivated in China for over a thousand years. It is grown in all regions of China except the Tibetan Plateau, which has the largest cultivation area and production volume in the world, accounting for over 45% of the total global yield. Watermelon is crisp, tender, juicy, sweet, and nutritious, with cooling and diuretic effects, making it a popular refreshing fruit-like vegetable during the summer.

I. Biological Characteristics

1. Morphological characteristics in botany

(1) Roots: The roots are deep and wide, with the taproot reaching a depth of over 1 m. When sown directly in sandy loam soil, the lateral roots can spread up to 3 m in diameter. The majority of the root system is distributed within the 20–30 cm of plow layer. The roots are thin, fragile, and have weak regenerative abilities, so they are not suitable for transplantation. With strong water and nutrient absorption capabilities, they are drought-tolerant but intolerant to waterlogging.

(2) Stems: The stems are upright during the seedling stage, with shortened internodes. After 4–5 nodes emerge, the internodes elongate, and crawling growth begins when the 5–6 leaves emerge. Watermelon plants have strong branching capabilities, forming 3–4 levels of lateral branches. There are also bushy-type watermelons with short internodes.

(3) Leaves: The leaves are alternate, with variations in parted leaf, lobate leaf, and leaf with entire margins. They are arranged in a 2/5 phyllotaxis. The leaf surfaces have fine hairs and wax powders, with low transpiration rates and drought tolerance.

(4) Flowers: The flowers are unisexual and axillary. Pistillate and staminate flowers blossom on the same plant. Staminate flowers appear at 3–5 nodes of the main stem, while pistillate flowers develop at 5–7 nodes, alternating with staminate

flowers thereafter. Bisexual flowers can also be present during the flowering period. The flower has 5 petals, with a yellow corolla that is fused at the base. The three anthers are fused together, with short filaments and split anthers. The inferior ovary is topped with a three-lobed stigma, forming a tricarpellary, three-chambered structure with parietal placentation. Both pistillate and staminate flowers have nectaries and are insect-pollinated. They open in the early morning and close in the afternoon.

(5) Fruits: Watermelon fruits come in various shapes, including round, oval, elliptical, and cylindrical. The weight of a single melon can range from 10–15 kg, with smaller ones weighing 1–2 kg. The peel can be smooth, with skin colors ranging from greenish-white, green, dark green, and blackish-green to black, often with fine mesh patterns or stripes. The flesh can be divided into two types: firm and watery, with various colors such as creamy white, light yellow, deep yellow, light red, or dark red. The soluble solid content of the flesh is generally around 10%–12%. The fine hairs on the pericarp disappear, and the rind becomes firm and thick. In the powdery type of fruit, the skin is covered in white powder, gradually transitioning from green to yellowish-green in color. In the dark green type, the skin color is a darker shade of green. During harvest, it is important to leave the fruit stalk intact and prevent collision and compression to facilitate storage.

(6) Seeds: The seeds are flat, ovate, or long ovate in shape, with a smooth or cracked surface. The seed coat can be brown, dark brown, or brownish-black, and can be monochrome or multicolored. The weight of 1,000 seeds varies depending on the seed type: large-seeded types ranging from 100–150 g, medium-seeded types from 40–60 g, and small-seeded types only from 20–25 g. Watermelons for seed production can reach a weight of 150–200 g.

2. Requirements for environmental conditions

(1) Temperature: Watermelons thrive in hot and dry conditions and have strong heat tolerance. The optimal temperature for growth is 25–30°C, with the strongest assimilation process at 30°C. A significant diurnal temperature difference (8–14°C) promotes fruit enlargement and sugar accumulation. They are extremely sensitive to cold, with plant damage occurring at 0–5°C and growth ceasing at 10°C.

The optimal temperature for seed germination is 25–30°C, with very few seeds germinating below 15°C or above 40°C. The optimal temperature for flowering and fruit set is 25–32°C, with poor development below 18°C. The best temperature for fruit enlargement and ripening is around 30°C.

(2) Light: Watermelons thrive in full sunlight and are sensitive to shade. They require at least 10–12 hours of sunlight per day for optimal growth. However, short daylight periods during the seedling stage promote the formation of pistillate flowers. They have a high requirement for light intensity, with a light saturation point of 80 klx and a compensation point of 4 klx.

(3) Water: Due to their rapid growth, large fruit size, and high yields, watermelons have high water requirements. However, watermelons have a strong root absorption capacity, allowing them to absorb water from deeper soil layers. Additionally, they possess drought-tolerant leaves, making them relatively resistant to dry conditions. They are extremely intolerant to waterlogging, and excessive water will cause the entire plant to suffocate and die. Watermelons also prefer dry air, with a suitable relative humidity of 50%–60% in the air.

(4) Soil and nutrition: Watermelon roots require good aeration, with loose sandy loam soil ideal for their growth. They can grow normally in soil with a pH range of 5–7. Adequate nitrogen and potassium fertilizers are needed for stem and leaf growth, while potassium fertilizer is especially important for fruit development. Additional application of phosphorus and potassium fertilizers can help enhance the sugar content of fruits. Watermelons absorb nitrogen, phosphorus, and potassium in a ratio of about 3.28 : 1 : 4.33. Additionally, grafted watermelons have a high demand for magnesium, as insufficient supply can lead to leaf blight.

II. Types and Varieties

Based on the different cultivation purposes, the watermelons can be divided into two categories: watermelons for seed production and for consumption, with the latter being the main cultivation type. Watermelons for seed production are also called "seeding-watermelons". Watermelons for consumption can be classified based on various factors. For example, by fruit size, they can be divided into four categories: small (below 2.5 kg), medium (2.5–5.0 kg), large (5.5–10 kg), and extra-large (above 10 kg) watermelons; by maturation periods, they can be divided into three categories: early-maturing, medium-maturing, and late-maturing varieties. There are also seeded and seedless watermelon varieties. Classification can also be based on fruit shape, flesh color, etc.

Currently, the main cultivated varieties in production include the following.

(1) Red-fleshed seeded varieties: Shoushan, Zhemi No. 3, Zhemi No. 4, Qingfeng, Heizhenzhu, Shouxing, Baolong, Baoguan, Heimeiren, Jinmeiren, Zaochun Hongyu, Hongxiaoyu, and Nabite.

(2) Orange-fleshed seeded varieties: Xinlan, Tianfeng, Chenglan, and Chengfeng.

(3) Yellow-fleshed seeded varieties: Jiahua, Heping, Baofeng, Xinjinlan, Texiaofeng, Xiaolan, Huangxiaoyu, etc.

(4) Seedless varieties: Ruilan, Nongyouxin No. 1, Xinghui, Lijing, Jinfeng, Jinshan, Haiwang, etc.

III. Cultivation Season and Methods

Watermelons require high temperatures, and the period from late April to August is generally the most suitable season for their growth in the Yangtze River basin (Table 1-2). Watermelons are generally sown before the last frost and emerge after the last frost period. Transplanting is recommended to be done after the last frost.

Table 1-2 Cultivation Season and Cropping Schedule for Watermelons in the Yangtze River Basin

Season and Cropping Schedule	Sowing Phase	Transplanting Phase	Harvest Phase
Spring Crop	December – February	Late January – Mid-March	Late April – Mid - June
Summer Crop	May	June	Late July – Early September
Autumn Crop	Late July – Early August	August	Early September – Late October

Watermelons should not be grown continuously in the same plot and should be rotated with other vegetables every 7–8 years. When watermelons are cultivated for continuous or rotational cropping in facilities, measures for grafting seedlings should be taken to prevent the occurrence of fusarium wilt.

IV. Cultivation Techniques

1. Variety selection

When targeting local markets, it is advisable to opt for early-maturing varieties; medium-maturing varieties are more suitable for export-oriented production; late-maturing varieties tend to set fruit late and have lower yields, making them unsuitable for greenhouse cultivation during spring cropping.

2. Grafting seedlings

Watermelon grafting cultivation primarily aims to prevent the transmission of soil-borne diseases, particularly fusarium wilt, to watermelon plants. It exhibits notable disease resistance and yield-enhancing effects while reducing the duration of crop rotation for watermelons. Yunnan black-seeded pumpkin, bottle gourd, and winter melon are used as rootstocks. Winter melon has poor tolerance to low temperatures, so it is mainly used for the grafting cultivation of watermelon in summer and autumn. Although pumpkin rootstocks have better tolerance to low temperatures than winter melon, they tend to cause excessive vine growth, delayed fruiting, and poor appearance and quality of the fruit. Currently, China has cultivated a batch of dedicated watermelon rootstocks, such as Chaofeng F_1, Nanza No. 8, and Xizhen No. 1, which are available for selection. The grafting method is similar to cucumber grafting, and both approach grafting and splice grafting can be adopted. After grafting, it is important to pay attention to moisture retention, insulation, and shading.

3. Land preparation and ridge tillage

Select loose sandy loam soil with high terrain, good drainage, and a deep soil layer for cultivation. Watermelons prefer well-drained soil and convenient irrigation, thrive in sunny conditions, and are drought-tolerant, but intolerant of waterlogging. Generally, a 3-year crop rotation is implemented. However, in recent years, measures such as planting disease-resistant varieties and applying agents for continuous cropping have allowed for certain levels of continuous cropping cultivation based on autumn plowing. For high ridge cultivation, the ridge width should be 4–4.5 m or 2–2.5 m. For medium-fertility soil, 2,000–2,700 kg of well-rotted manure, 50–100 kg of microbial complex fertilizer, and 67–130 kg of cake fertilizer should be applied per 1,000 m^2. The base fertilizer for early-maturing cultivation should be increased by 60%–70%, and the full-layer fertilization method should be employed when there is sufficient organic fertilizer.

4. Transplanting

When the soil temperature at a depth of 10 cm remains stable above 14°C, and the air temperature in the greenhouse remains stable above 10°C, it can be regarded as a safe transplanting period. According to the climate characteristics of "three colds and four warms" in early spring, it is recommended to select a sunny morning for transplanting when the cold fronts have ended and the warm fronts are approaching. The suitable transplanting density is as follows.

(1) For crawling cultivation, early-maturing varieties can be planted at a row spacing of 1.6–1.8 m or wider spacing of 2.8–3.2 m and plant spacing of 40 cm, with approximately 1,000 seedlings per mu. For medium-maturing varieties, it is recommended to have row spacing of 1.8–2 m or wider spacing of 3.4–3.8 m, with a plant spacing of 50 cm and approximately 800 seedlings per mu.

(2) For trellis or suspension cultivation, the plants can be planted at the row spacing of 1–1.2 m and the plant spacing of 40 cm for early-maturing varieties and 50 cm for medium-maturing varieties, with about 1,350–1,500 seedlings per mu.

When grafted seedlings are planted, it is important to ensure that the seedlings are planted shallowly, with the planting depth aligning with the soil clump. The graft union should be kept above the ground level during soil covering to prevent the formation of adventitious roots that could potentially affect the grafting effectiveness. It is recommended to plant seedlings of different sizes in separate areas. Larger seedlings should be planted on both sides of the greenhouse, while smaller ones should be planted in the center of the greenhouse. This arrangement facilitates better management of the seedlings. After planting, the planting furrows should be filled with water to ensure that the water penetrates the soil clump and surrounding soil. It is important to ensure sufficient water during transplanting and to cover the furrows with soil after watering.

5. Field management

(1) Fertilizer and water management.

The principle of top dressing is to apply less amounts of fertilizer for the early growth of seedlings, skillfully apply fertilizer for vine growth, and apply large amounts of fertilizer for fruiting. Under the condition of sufficient base fertilizer, no top dressing is required before the fruit set. The first application should be made after the fruit has set, followed by another application one week later. The fertilizers should mainly consist of nitrogen and potassium, with phosphorus fertilizer applied as a supplementary top dressing. In the case of staggered fruiting, as the fruiting period prolongs, it is essential to increase the amount of top dressing to meet the nutritional requirements for the continued growth of vines and leaves, as well as sustained fruit development.

Regarding water management, aside from applying top dressing and watering after the fruit set, irrigation is generally done only under dry soil conditions. Watering is ceased 7–10 days prior to harvest.

(2) Intertillage, weeding, and soil cultivation.

Shallow intertillage is recommended. Soil cultivation should be carried out before the plants start to grow, in conjunction with intertillage. Soil cultivation can prevent plants from being blown by the wind and ensure their smooth growth.

(3) Plant regulation.

There are three main methods for training watermelon vines: ① Single vine system: Only the main vine is retained, and all side shoots are removed. ② Double vine system: In addition to the main vine, one vigorous side shoot is selected between the 3rd and 4th nodes of the lower part of the plant, and all other lateral vines are removed. ③ Triple vine system: In addition to the main vine, two vigorous side shoots are selected between the 3rd and 5th nodes of the lower part of the plant, and all other lateral vines are removed. The training of vines should not be done too early, but should be typically carried out after the vines have started to crawl and produce lateral vines. During training, both the main vine and the selected side shoots should be guided in the same direction while maintaining a suitable distance from each other to allow for forward growth and overlapping of the vines.

When grafting watermelons, it is crucial to fix the vine without burying it underground. Otherwise, if the watermelon stem and vine root into the soil after being covered, the grafting process will become meaningless.

When the vine reaches a length of around 50 cm, preferably on a sunny and warm afternoon, carefully train the watermelon vine across the furrows to the adjacent high ridge. Secure it in place using slender twigs or branches, ensuring that the watermelon seedling grows in the desired direction. The main vine and lateral vines can be directed in the same or opposite directions, and the distribution of the vines should be uniform.

(4) Hand pollination and fruit thinning.

During the flowering and fruiting period, gently rub the stigma of the flowers from 6:00 to 10:00 each day, ensuring that the pollen is evenly transferred to the stigma. Generally, one staminate flower can pollinate three pistillate flowers. After pollination, hang a tag on the node of the pollinated flower, indicating the date of pollination for future reference during harvest. To ensure a good fruit setting rate, pollination should be performed on the first 1–3 pistillate flowers on the main vine and the first pistillate flower on each retained lateral vine.

When the fruits reach the size of an egg, they should be thinned. The sequence of fruit thinning is as follows: Select 1 fruit from the second and third fruits on the main vine. If the fruit on the main vine is not suitable for leaving on the vine due to its failure to set or poor quality, select a fruit from the lateral vine to keep, with only 1 fruit per plant to be kept. Choose fruits that are well-shaped and consistent with the characteristics of the variety, have bright-colored skin, and grow fast. The rest should be thinned from the fruit stalk with scissors.

Generally, the seedlings of the second crop should be pollinated when the fruits of the first crop reach the desired size. Early pollination may lead to intense competition for nutrients between the two crops of fruits, hindering the normal growth and timely maturity of the first crop of fruits. Late pollination may delay the marketing of the second crop of fruits, resulting in lower cultivation efficiency. Fruit thinning should be carried out early for the second crop, regardless of their position on the plant. The first fruit that sets should be left, and the rest should be removed.

(5) Fruit management.

① Padding the melon: After the young melons shed their hairs, pad the melons with straw rings that are made of clean wheat straw or rice straw, lifting them off the ground to maintain good ventilation and prevent damage from soil-borne pathogens and underground pests.

② Turning over the melon: The main purpose of turning the melons is to expose the entire surface to light and achieve even coloring. Melon turning generally starts on a sunny and warm afternoon after the fruits are selected. Gently lift the melon with both hands and slowly rotate it in one direction so that approximately half of the shaded portion is lifted off the ground. Rotate the melon 2–3 times to position the shaded side towards the sun. When the melons are turned over, they should be rotated in the same direction.

③ Standing the melon: The main purpose of standing the melon is to adjust the size of the melons and ensure the symmetry of the upper and lower parts. During the fruit growth period, for melons with significant differences in thickness between the two parts, stand the melons with the thinner end downwards and the thicker end upwards, resting the lower end on the straw ring.

④ Supporting or hanging the melon: When the watermelon grown on trellises reaches around 500 g (after thinning), support it from below with a net bag or straw ring.

6. Harvesting

(1) Criteria for ripe melons.

The ripeness of a fruit can be determined based on the following aspects.

1) Changes in appearance.

① Changes in tendrils. In general, when the tendrils on the nodes where the melon is attached and the 1–2 nodes before and after the melon turn yellow or wither, it indicates that the melon on that node is ripe. However, for the currently cultivated high-quality small early-maturing watermelons, the tendrils on the nodes where the melon is attached and the 1–2 nodes after the melon generally do not turn yellow or wither when the melon is ripe. Once they turn yellow or wither, it indicates that the melon is overripe.

② Changes in fruits. When a watermelon is ripe, its rind becomes visibly brighter and harder. The base color and pattern of the rind exhibit a clear contrast with distinct patterns and edges, indicating signs of aging. Watermelons with ribs will have prominent and well-defined ridges. Scratched areas and the stem end of the fruit show evident inward depressions. The stem becomes twisted and aged, and the fine hairs at the base tend to fall off.

2) By date.

This method is relatively accurate with minimal error and is most suitable for the cultivation of watermelons in a controlled environment. The basis for this method is as follows: for watermelons cultivated in greenhouses in early spring, from the blossom of pistillate flowers to the ripening of the fruits, early- to medium-maturing varieties generally require 28–35 days, while medium- to late-maturing varieties require around 40 days. For the same variety, the first crop usually takes an additional 3–5 days compared to the second crop.

Once the fruits have reached the required number of days, select the representative melons from the same batch, and cut them to check if the actual ripeness matches the determined ripeness. If they match, the melons of the same date can be harvested for market.

3) Changes in sound.

When you tap on the melon with your finger, a low "thud" indicates ripeness, while a crisp "crackling" sound indicates immaturity. Ripe watermelons with thin pericarp are prone to cracking, so you should be careful when tapping.

(2) Harvesting time and method.

Harvesting in the morning when the melons are cool makes them easier to preserve, with higher water content and juice, better taste, more freshness, and higher yields. When harvesting, cut off the vine of the melon with scissors, including 1–2 nodes before and after the node where the melon is attached, allowing the watermelon to have a stem with a section of vine and 1–2 leaves.

7. Common cultivation issues and prevention strategies

When cultivating early-maturing watermelons in spring, common issues, such as emergence, with seed coat, uneven emergence, seedlings without growing points, elongated seedlings, and stunted seedlings, often occur during the seedling stage. The emergence of seed coat is caused by insufficient watering and inadequate soil covering. It can be prevented by providing sufficient water, covering the seed with the soil of 1.5 cm, and gently removing the seed coat in the moist morning of the seedbed. Uneven emergence is mainly due to uneven temperature and humidity in the seedbed. Seedlings without growing points are caused by sowing old seeds or excessive application of fertilizer or pesticides. Watermelon seedlings with no growing point are more common in seeds stored for over three years. Newly emerged seedlings have tender growth points and are less tolerant to fertilizers and pesticides. Spraying pesticides or applying high concentrations of fertilizer in large amounts can easily damage the growth point. The main causes of elongated seedlings are excessively high humidity and temperature in the seedbed (especially high nighttime temperatures), excessive application of readily available nitrogen fertilizer, insufficient sunlight, and small spacing between seedlings. It is necessary to prepare nutrient-rich soil as required, strengthen the management of the temperature and humidity of the seedbed, ensure that the nighttime temperature after emergence does not exceed 15°C, water properly, enhance the ventilation, lower the humidity, increase light exposure in the seedbed, and adjust the spacing between seedlings. Stunted seedlings are characterized by small leaves, dark color, thin stems, short internodes, slow growth, and few roots. They are mainly caused by prolonged low temperature or dryness in the seedbed, insufficient fertilization, lack of nitrogen, or excessive fertilizer application. To prevent stunted seedlings, nutrient-rich soil should be used for seedling cultivation to avoid nutrient deficiency and root burning. Maintain suitable temperature and humidity in the seedbed and carry out proper seedling hardening according to the seedling condition and weather conditions.

「**Summary of Sub-context**」

In watermelon production, it is important to first select high-quality varieties. Watermelons are highly intolerant to waterlogging, so the planting area should not accumulate water to avoid vine and fruit rot. After a batch of melons is harvested, timely additional top dressing should be conducted. Moreover, it is recommended to harvest the melons after the morning dew has dried, as sufficient moisture in the morning can easily cause melons to burst when tapped.

「**Expanded Knowledge**」

The Origin of Seedless Watermelons

Common watermelons are diploid plants, which means that they have two sets of chromosomes (2N=22). After the seedlings are treated with colchicine, the chromosomes of the diploid watermelon plant are doubled to become tetraploid (4N=44). Such tetraploid watermelon can also produce flowers and set fruits, and its seeds can germinate and grow normally. Then, the tetraploid watermelon plant is used as the maternal parent (with the staminate flowers removed), and a diploid watermelon plant is used as the paternal parent (with its pollen used to pollinate the tetraploid pistils) for hybridization. In this way, triploid seeds can be obtained from the tetraploid watermelon plant. When the triploid plant blossoms, its pistils should be pollinated by the normal diploid watermelon pollen to stimulate the fruit development from the ovary. The ovules cannot develop into seeds, but the fruit can develop normally, which thus results in a seedless watermelon!

Sub-context 3　Production of Netted Muskmelon

Muskmelon cultivation in China has a long history. Muskmelon is loved by people for its sweet and delicious taste. Especially, the highly evolved netted muskmelon has entered modern consumer society as a high-end fruit. As a year-round fresh fruit, it is favored by people from all walks of life and is sold well at high prices both domestically and internationally. In the regions south of the Yangtze River in China, netted muskmelons are cultivated in greenhouses with high economic and social benefits.

I. Biological Characteristics

1. Botanical features

(1) Roots: The roots are well-developed and penetrate deep into the soil, with the main root system distributed within 30 cm of the soil layer. They undergo cork formation easily and have poor root regeneration ability, so it is advisable to sow the seeds directly or take protective measures for seedling cultivation.

(2) Stems: The stem is hollow and has stiff hairs, and lateral branches can grow at each internode. Lateral branches that grow on the main vine are called side shoots, and lateral branches that grow on the side shoots are called secondary shoots. For thick-skinned muskmelons, fruit setting mainly occurs on the side shoots.

(3) Leaves: There are two cotyledons, elongated elliptical in shape. True leaves are nearly circular or kidney-shaped, with entire or 5 lobed margins. They are green or dark green, alternate on the stem, and covered with fine hairs.

(4) Flowers: Unisexual flowers, monoecious. Staminate flowers are smaller and clustered. Pistillate flowers have inferior ovaries and are solitary, without parthenocarpy ability. Pistillate flowers on the main vine appear late, while those generally appear at 1–2 nodes on the lateral vines. They are insect-pollinated flowers, making them cross-pollinated. The flowers have a short blooming period, starting from 5:00 to 6:00 in the morning and withering in the afternoon.

(5) Fruits: The fruit is pepo. The shape, size, and color of the fruits, as well as the pattern on the pericarp, vary greatly depending on the variety, which is considered as one of the main criteria for variety identification. The flesh of the fruit is soft or crisp with fragrance.

(6) Seeds: The seeds are smaller than cucumber seeds and show significant

differences in size. The seeds have a lifespan of 3 years.

2. Requirements for environmental conditions

The environmental requirements for muskmelons are roughly the same as for watermelons. They prefer warmth and sunlight, are drought-tolerant but intolerant of waterlogging, and require a lot of nutrients and water during the fruit expansion period.

(1) Temperature: Muskmelons prefer warmth and are sensitive to cold. The optimal temperature for the germination period is 25–35°C, and the suitable temperature for growth is 22–32°C. Poor growth occurs below 12°C. Muskmelons have strong heat tolerance and can withstand temperatures above 35°C.

(2) Light: Muskmelons prefer light and are intolerant of shade. The light compensation point is 4 klx, and the saturation point is 70–80 klx. During the fruiting period, they require more than 10–12 hours of sunlight, and poor fruiting occurs with less than 8 hours of sunlight.

(3) Water: Muskmelons have a strong tolerance to dryness and drought. The suitable air humidity is 50%–60%, and during the flowering and fruit setting period, the relative air humidity is required to be around 80%. Excessive soil humidity is prone to root rot.

(4) Soil and nutrition: Muskmelons are not strict with soil requirements and have strong adaptability. Sandy loam soil with deep, loose, and well-aerated soil layer is the best. The suitable soil pH is 6.0–6.8. Muskmelons prefer phosphorus and potassium fertilizers, and have a relatively high demand for calcium, magnesium, and boron. They have moderate salt tolerance ability.

II. Varieties

Xinjiang, Gansu, and other regions are the main cultivation areas for thick-skinned muskmelons in China. Since the 1990s, other regions in China have also achieved good results by cultivation in greenhouses and polytunnels. Therefore, the cultivation scale has expanded rapidly. Representative varieties of smooth-skinned melons include Elizabeth, Zhengtian No.1, Zhuangyuan, Mishijie, Milu, Tianmi, and Yulu. Netted muskmelons include Daqing Migua, Tianmi, Huaguan, etc.

III. Cultivation Season and Methods

In northern regions, netted muskmelons are mainly cultivated in a controlled environment, such as energy-saving sunlight greenhouse and spring-autumn greenhouse cultivation. The Cultivation season and cropping schedule for netted muskmelons in the Yangtze River basin can be found in Table 1-3.

Table 1–3 Cultivation Season and Cropping Schedule for Netted Muskmelons in the Yangtze River Basin

Season and Cropping Schedule	Sowing Phase	Transplanting Phase	Harvest Phase
Spring Crop	Early February – Mid-February	Early March – Mid - March	Before the rainy season
Autumn Crop	Late July	Early August	Before the first frost

Muskmelons should not be continuously grown in the same plot and should be rotated with non-Cucurbitaceae vegetables at least every 4–5 years. Otherwise, grafting seedlings should be adopted.

IV. Cultivation Techniques

1. Variety selection

It is recommended to select excellent varieties that are well tolerant to low temperatures and have a concentrated maturity period, early maturity, and disease resistance, such as Elizabeth, Zhuangyuan, Huaguan, etc.

2. Seedling raising

Netted muskmelons have a relatively short period for seedling raising. It is suitable to raise seedlings in a greenhouse at the age of 30–35 days and transplant them when they have 3–4 true leaves. The nutrient bowls should be employed for root protection during the seedling raising. Melon seeds are prone to carrying viruses, so the seeds should be disinfected and soaked for germination before sowing. High temperature should be maintained during the germination period, and seedlings generally emerge 3 days after sowing. Watering should be controlled during the seedling raising period to prevent excessive elongation of the melon seedlings.

Grafting seedlings are commonly cultivated with wild melons and black-seeded pumpkins as rootstocks. The grafting method used is the splice-grafting technique. For specific instructions, please refer to the techniques for grafting seedlings.

3. Fertilization and ridge tillage

Apply formulated fertilizers, including 3,000–5,000 kg/mu of high-quality organic fertilizer, 50 kg/mu of compound fertilizer, about 50 kg/mu of calcium magnesium phosphate fertilizer, about 20 kg/mu of potassium sulfate (potassium chloride is prohibited), and 1 kg/mu of boron fertilizer. Small-sized melons have a high planting density, allowing for even fertilization. After fertilization, to perform deep soil tillage is recommended. Large-sized melons require centralized fertilization in the trenches due to their low planting density.

When high ridges or raised beds are employed for cultivation, the width of high ridges is 90–100 cm with a height of 15–20 cm, and two rows of seedlings are planted per ridge; the width of raised beds is 40–50 cm with a height of 15–20 cm, and one row of seedlings is planted per bed.

4. Transplanting and mulching film covering

Transplanting begins when the minimum temperature inside the greenhouse stabilizes above 5°C. 1,500 seedlings of large fruit plants should be planted per mu, with an average row spacing of 80 cm and a plant spacing of 50 cm. Approximately 1,800 seedlings of medium-sized fruit plants should be planted per mu, with an average row spacing of 80 cm and a plant spacing of 45 cm. About 2,100 seedlings of small fruit plants should be planted per mu, with an average row spacing of 80 cm and a plant spacing of 40 cm. Cover the plants with mulching film.

5. Field management

(1) Temperature management.

During the initial stages of transplanting, when the external temperature is low, row covers and double-layer shade cloth can be set up to promote seedling acclimation. The daytime temperature should be maintained between 25–30°C, and the nighttime temperature should not drop below 15°C. Netted muskmelons are sensitive to low temperatures during the formation of net patterns. When the temperature drops below 18°C, the hardening of the pericarp is delayed, resulting in sparse and coarse net patterns. After the melon seedlings survive, the temperature should be lowered to around 25°C during the day and 12–15°C at night. During the fruit development stage, the daytime temperature should be maintained at 25–30°C to promote photosynthesis. Gradually reduce the mulching and increase ventilation, allowing for greater temperature difference between day and night so as to aid in nutrient accumulation, increase sugar content, and improve the quality.

(2) Fertilizer and water management.

Sufficient water should be provided after transplanting. Generally, water is not supplied after mulching and before the fruit set, especially during the flowering and fruit setting period, to prevent excessive vine growth and flower drop. Watering should start after the fruit set, and the ground should always be kept moist to avoid alternating dry and wet soil, which can lead to dehiscent fruit. During the fruiting stage, ventilation should be enhanced to avoid excessively high humidity in the air. During the net pattern formation period of netted muskmelons, excessive air humidity can not only affect the quality of the net patterns but also lead to disease in the cracked areas of the melon surface.

(3) Plant regulation.

① Pruning: For muskmelons cultivated in greenhouses and polytunnels, the multiple-vine pruning method is commonly used. The specific approach is as follows: Select 2–3 side shoots with pistillate flowers before and after the node where the melon is attached as reserve fruiting vines. Leave 1–2 leaves before the first pistillate flower, and pinch all the other lateral vines.

② Hanging vines: When the melon vines grow, prepare a thin nylon rope or long strip of cloth for each melon plant. Plastic ties are not recommended to prevent injury to the melon if the ties break. Attach the upper end of the rope to a horizontal wire and loosely tie the lower end to the base of the melon seedling. As the melon vine grows, periodically wind the vine around the hanging rope.

(4) Hand pollination and fruit thinning.

In greenhouses, hand pollination is necessary due to the lack of insect pollinators. The specific method is as follows: pick freshly blooming staminate flowers from the plants before 10:00 on the day when the staminate flowers are blooming, remove the petals of staminate flowers to expose the stamen, and gently rub the anther on the stigma of the pistillate flowers to evenly transfer the pollen to the surface of the stigma. When doing hand pollination, sufficient and uniform pollination should be ensured to avoid malformed melons.

When the fruits reach the size of an egg, they should be selected and thinned. For small-fruited varieties, keep two melons per plant, while keep one melon per plant for large-fruited varieties. The appropriate positions for fruit setting are around the 12th–15th node for small-fruited varieties and the 15th–18th node for large-fruited varieties.

(5) Other management.

① Hanging and turning melons: For small-fruited varieties, when the melons reach the weight of 250 g, they can be individually hung in plastic mesh bags, horizontally positioned to prevent vine and melon from dropping. Additionally, after hanging the melons, turn them over 2–3 times at 13:00–14:00 regularly to ensure even exposure and coloration.

② Leaf picking and pinching: Old, diseased, and yellow leaves at the lower part of the plant should be removed early to reduce disease incidence and promote better ventilation and light penetration.

For small-fruited varieties, pinching should be done when the main vine has developed 20–25 leaves, while for large-fruited varieties, it should be done when the main vine has developed 25–30 leaves to promote early ripening of netted muskmelons.

6. Pest and disease control

Throughout the growth period, muskmelons are susceptible to diseases such as seedling blight, downy mildew, anthracnose, powdery mildew, and blight, so measures should be taken, and comprehensive control methods should be implemented.

7. Harvesting and post-ripening

The characteristic feature of mature fruits is the presence of brown spots on the leaves at the nodes of fruiting branch, and a distinct color, aroma, and netting pattern on the fruit skin, particularly in varieties. The netting pattern on the surface of the muskmelon should be clear, dry, and dark in color, and the pericarp should be firm. The stem may turn yellow or fall off on its own, and tendrils near the nodes where the melon is attached should be dry and withered. When tapping the melon with your palm, a ripe melon will sound dull. However, the correct harvest time should be determined based on the fruit setting date and the maturity of the variety. Harvesting in the morning, when the fruit has a high moisture content, results in good quality. Use scissors to cut the fruit with a small piece of stem in a "T" shape. After harvesting, muskmelons need a certain amount of time for post-ripening. Post-ripening causes the pericarp to become thinner, the fruit to become softer, and the taste to significantly improve, which is especially important for netted muskmelons as post-ripening generally takes 3–7 days.

「Summary of Sub-context」

Netted muskmelons are a high-end type of muskmelons at a high price with good quality. During cultivation, attention should be paid to fertilizer and water management, plant regulation, and comprehensive prevention and control of pests and diseases.

「Expanded Knowledge」

Standards for Netted Muskmelon Grading

Grade A: The fruit has a standard shape, clear and regular netting pattern, no scars, and a T-shaped stem.

Grade B: The fruit has a standard shape, clear and relatively regular netting pattern, 1–2 scars, and a T-shaped stem.

Grade C: The netting pattern is not clear, and the fruit may have deformities, scars, and incomplete stems.

「Reviewing and Thinking Questions」

I. Explanation of Terms

1. Fruit abortion of cucumber.
2. Emergence with seed coat.
3. Cucumber parthenocarpy.

II. Gap Filling

1. Common grafting methods for melon vegetables include ________, ________, and cleft grafting.

2. The common rootstocks used for cucumber grafting are ________; their functions are ________.

3. Muskmelons are mainly produced on ________, while early maturing cucumbers are mainly produced on ________. Zucchini are mainly produced from ________.

4. The common rootstocks used for watermelon grafting are ________.

5. Watermelon has a ________ ability to regenerate its root system and is tolerant to ________ but not resistant to ________.

6. Cucumbers should be harvested promptly when they are young, and in particular, ________ cucumbers should be harvested early.

7. Hand pollination of watermelons should be performed at ________.

8. Melon vegetables that commonly produce their fruits on main vine include ________, ________, etc.; melon vegetables that can bear fruit on both main and lateral vines include ________, ________, etc.; melon vegetables that mainly bear fruit on lateral vines include ________, etc.

9. Common diseases in watermelon production include ________, ________, ________, ________, etc. Common pests include ________, ________, etc.

10. Watermelon vine pressing can promote ________ and ________ of plants.

III. Choice Questions

1. (　　) and (　　) promote the differentiation of pistillate flowers in

cucumbers.

A. Low nighttime temperature, short daylight

B. Low nighttime temperature, long daylight

C. High nighttime temperature, short daylight

D. High nighttime temperature, short daylight

2. During which period does water and nutrient absorption of watermelon reach its peak? (　　).

A. Fruit-setting period　　B. Fruit-expanding period

C. Germination period　　D. Ripening period

3. In hybrid seed production of which vegetable, the staminate flowers are directly removed? (　　).

A. Cucumber　　B. Tomato　　C. Kidney bean　　D. Cabbage

4. Cucumber's downy mildew is a type of (　　).

A. Fungal virus　　B. Bacterial pathogen

C. Viral disease　　D. Physiological disease

5. The melon vegetable which has least stress tolerance is (　　).

A. Pumpkin　　B. Gourd　　C. Cucumber　　D. Winter melon

6. The time for hand pollination of watermelons is generally at (　　).

A. 6:00–9:00　　B. Around 12:00

C. Afternoon　　D. 15:00–16:00

7. The period when watermelon and muskmelons require sufficient irrigation is (　　).

A. Flowering period　　B. Fruit-expanding period

C. Stabilization period　　D. Harvest period

8. The sex type of cucumber flowers is controlled by hormones; (　　) can increase the number of pistillate flowers.

A. Gibberellin　　B. Chlormequat

C. Ethephon　　D. Abscisic acid

9. Which of the following crops is not suitable to serve as a rootstock for grafting watermelons? (　　)

A. Wild watermelon　　B. Cucumber

C. Gourd for consumption D. Black-seeded pumpkin

10. The melon vegetable which has the least stress tolerance is ( ).

A. Pumpkin B. Gourd C. Cucumber D. Winter melon

11. Which method of vine pressing is preferred for grafting watermelons? ( )

A. Fixation without soil covering B. Fixation by soil covering

C. Arbitrary fixation D. No fixation

IV. Thinking and Answering

1. What causes deformed cucumbers like Big-belly Cucumber, Narrow-waisted Cucumber, and Pointy Cucumber in cucumber cultivation?

2. How to determine the ripeness of watermelons?

3. What factors influence the sexual differentiation in cucumbers?

4. What are the advantages of grafting seedlings?

Learning Context 2 Production of Solanaceous Vegetables

「**Learning Objectives in This Context**」

Master the main types and common characteristics of solanaceous vegetables, the biological characteristics of the main types, and basic knowledge of solanaceous vegetable production. Be able to choose the appropriate varieties of solanaceous vegetables based on cultivation facilities and seasons. Be capable of creating production plans for solanaceous vegetables and carrying out production correctly.

「**Analysis of Tasks in This Context**」

Seedling cultivation and grafting, plant regulation, pollination, flower and fruit retention, fertilization and irrigation management, and pest and disease control in solanaceous vegetable production.

「**Introduction**」

In our daily life, solanaceous vegetables are one of the most frequently consumed vegetables. They play an important role in people's dietary structure, as they have rich nutrition and significant economic benefits.

Overview

I. Types of Solanaceous Vegetables

Solanaceous vegetables include tomatoes, eggplants, pepinos, chili peppers, wolfberries, and tree tomatoes (Fig. 2-1). They all belong to the Solanaceae family. The edible parts are berries. Tomatoes, eggplants, and peppers are widely cultivated throughout China due to their strong adaptability, high yields, and long supply season. They are suitable for cultivation both in open fields and in a controlled environment.

Fig. 2-1 Different Solanaceous Vegetables

Wolfberries are mainly used for medicinal purposes, and in Guangdong, Guangxi, and other areas, their tender leaves are also consumed as vegetables. They are less cultivated in other regions. Chinese lanterns have limited cultivation as well. Pepinos are commonly consumed as fruits and vegetables in countries such as Colombia, Ecuador, Peru, and Chile. They have beautiful colors with light yellow and purple stripes, and a fragrance similar to muskmelons, but they are not excessively sweet. Fresh consumption is the main method, and there has been limited variety induction in many parts of China. Tree tomatoes are perennial evergreen or deciduous shrubs. They are native to South America and were introduced to the UK as ornamental plants. They are widely distributed in the southern United States, and there are small populations in Yunnan Province, China. Their taste is similar to tomatoes but extremely sour, making them unsuitable for direct consumption. They can be used for cooking or processed into various products.

II. Characteristics of Solanaceous Vegetables

The most commonly cultivated solanaceous vegetables in Jiangsu Province are tomatoes, eggplants, and chili peppers. They are widely grown in urban suburbs and rural areas, and become the main spring vegetables and seasonings.

In addition to being consumed freshly, solanaceous vegetables are also good materials for processing. Tomatoes can be used to make tomato juice, tomato sauce, and dried tomatoes. Eggplants can be dried to make dried eggplants. Chili peppers can be made into chili sauce or chili powder. In southwestern and northwestern China, as well as Hunan, Jiangxi, etc., some peppers are produced as dried chili, while others are consumed as fresh vegetables.

The main nutritional value of solanaceous plants lies in their rich content of vitamins, mineral salts, carbohydrates, organic acids, and a small amount of protein, all of which are essential nutrients for the human body. Both chili peppers and tomatoes are rich in vitamin C. Fresh chili peppers contain 75–185 mg of vitamin C per 100 g, and the content is even higher when they are fully ripe. On the other hand, tomatoes generally contain 20–30 mg of vitamin C. However, the content of ascorbic acid in eggplants is not high, and the various nutrients are greatly influenced by variety selection and cultivation techniques.

The cultivation of solanaceous plants has the following characteristics.

(1) Wide adaptability. Although solanaceous plants are short-day plants, they are not sensitive to photoperiod. As long as the temperature is suitable, they can bloom and bear fruit. However, tomatoes do not grow well in excessively hot summers, so autumn and winter cultivation is the main practice in Guangdong and Guangxi. In Yunnan, chili peppers can be grown in open fields during winter.

Solanaceous vegetables also have strong adaptability to different types of soil, but they prefer well-drained, fertile sandy loam soil. In various regions of South China, where rainfall is abundant, deep furrows and high ridges are used to facilitate drainage, increase soil temperature, and promote the development of the root system.

(2) High nutrient demand. Solanaceous plants have a long growing season and a long fruiting period, with multiple harvests (for processing varieties, they can be mechanically harvested at once), which requires a large amount of fertilizer and water. Adequate nitrogen fertilizers, as well as a certain amount of phosphorus and potassium fertilizers, are needed as base fertilizers. Early and multiple applications of top dressing are also necessary. For early-maturing varieties, early top dressing is particularly important.

(3) Strong seedling cultivation. In the Yangtze River basin, for the purposes of early supply and extension of the growth and fruiting periods, tomatoes, eggplants, and chili peppers are often cultivated in multi-layered greenhouses. The cultivation of strong seedlings is a characteristic of solanaceous plant cultivation. However, when transplanting, the seedlings should not be too large.

(4) Increase in planting density. Increased yield through close planting is mainly due to an increase in the number of fruits, with a particularly significant increase in early yields. However, if the planting density is too high, the weight per plant and the weight per fruit tend to decrease, leading to the failure in high yields.

(5) Plant regulation. Chili pepper and eggplant plants are shorter with upright stems, and their branching is less or not pruned. Tomatoes, on the other hand, have trailing stems. In various southern regions, trellises are used for cultivation, and pruning and pinching are carried out. If they are grown on the ground without trellis, the fruits may come into contact with the soil, making them susceptible to rot or insect damage. Using trellises, pruning, and pinching can also increase the planting density. However, they are cultivated without trellis in some areas and for processing varieties.

(6) Serious diseases and pests. Most diseases and pests of solanaceous plants are parasitic, including bacterial wilt, early blight, late blight, brown spot, and various viral diseases. There are also various physiological disorders, such as

leaf curl, stem breakage, blossom end rot, internal browning (also caused by viral diseases), dehiscent fruit, sunscald, and misshapen fruits, which not only affect fruit yield but also fruit quality. These are the main issues in solanaceous plant cultivation.

Sub-context 1 Production of Tomato

Tomato, also known as *Solanum lycopersicum*, is an annual vegetable belonging to the *Solanum* in Solanaceae family. It is native to South and Central America, with wild varieties found in the Andes Mountains. Tomato cultivation was introduced to Europe from Mexico in the 16th century. Initially, it was grown as an ornamental plant, and only in the second half of the 18th century did it start being used as a vegetable for consumption. Tomatoes were introduced to China around the late 16th to early 17th century.

I. Biological Characteristics

1. Botanical features

Tomato is a member of the Solanaceae family and is native to tropical regions of South America. It is a perennial plant in tropical climates but is grown as an annual crop in temperate regions.

(1) Roots: Tomatoes have a deep and extensive root system that reaches up to 2 m long. However, 60% of the root system is distributed within the top 30 cm of the soil, known as the plow layer. During transplantation, the damage of taproot may promote the growth of lateral roots and adventitious roots, increasing the root system within the 30 cm plow layer to over 85%.

(2) Stems: Tomato stems are semi-upright or trailing stems in nature. During the seedling stage, the plants grow upright without branching. The mature plants develop sympodial branching. After the appearance of the terminal inflorescence, due to apical dominance, the first lateral branch below it exhibits the strongest growth vigor. It replaces the main stem to continue growing, which causes the

terminal inflorescence to become a lateral one, and subsequently, lateral branches are formed in the same manner. Such botanical characteristic of tomatoes is known as sympodial branching or pseudo-sympodial branching in botany (Fig. 2-2).

Fig. 2-2 Botanical Characteristics of Tomatoes

Based on the growth characteristics of the main stem and the flowering habits, tomato plants can be classified into two types: indeterminate growth and determinate growth.

(3) Leaves: The leaves of tomatoes are large, which resemble compound leaves. Actually, they are single leaves composed of multiple unevenly sized lobes formed by deep incisions and partitions (some also consider them as compound leaves). The leaf shape can be ordinary, wrinkled, or potato-like, with colors ranging from deep green to light green or yellow. The stems and leaves of tomatoes are covered with short, fine hairs and secrete a distinctive odoriferous fluid. The features, including leaf shape, size, incisions, density, color, etc., are important criteria for distinguishing tomato varieties.

The stems and leaves of tomatoes are covered with glandular hairs that secrete fluids, emitting a distinct aroma. These hairs play a role in repelling certain insects, especially aphids. Therefore, the quantity of these glandular hairs is somewhat related to the degree of pest damage.

(4) Flowers: The flowers are arranged in cyme, and small-fruited varieties have a raceme. There are typically 5–6 floral organs, with 6 being more common. The inflorescence emerges at the internodes, and the number of flowers per inflorescence varies greatly among different varieties, from 5–10 flowers. Tomatoes are self-pollinating, but some varieties can undergo cross-pollination when influenced by the environment. The stigma extends out of the anther tube, allowing for cross-pollination. The natural hybridization rate ranges from 4%–10%. Flowers formed under unfavorable conditions, such as low temperatures, have swollen and flattened stigma, which may result in the development of misshapen fruit.

The floral buds of tomatoes differentiate quite early. Usually, when the seedlings have 2–3 true leaves, the first inflorescence begins to differentiate. The flowering and fruiting habits of tomatoes can be divided into two categories based on the positions of the inflorescence and the growth characteristics of the main stem.

① Determinate growth. After the main stem grows 6–8 true leaves, it starts to bear the first inflorescence, with one inflorescence every 1–2 leaves thereafter. After the main stem bears 2–4 inflorescences, the apical bud differentiates into an inflorescence and the stem no longer extends, resulting in topping. Therefore, the plants are short, with early and concentrated flowering and fruiting, a short supply period, and a high yield in the early stage, so they are suitable for early-maturing cultivation, such as Hezuo 903, etc.

② Indeterminate growth. After the main stem grows 7–9 true leaves, it starts to produce the first inflorescence, followed by one inflorescence every 2–3 leaves, and the top of the main stem can continue to grow upwards. The lateral branches that emerge from the leaf axil can also produce inflorescences in the same way. Therefore, this type of plant is tall and can produce 7–8 or more inflorescences on the main stem, bearing 7–8 clusters of fruits. It has long flowering and fruiting periods, and high total yield, such as Maofen 802, etc.

However, the fruiting habit is not always fixed. Some varieties have fruiting habits that fall between determinate and indeterminate growth, such as Zaoquezuan, which has determinate growth with an indeterminate growth tendency. Due to the differences in fruiting habits, there are significant differences in cultivation distance, pruning methods, maturity timing, and yield.

(5) Fruits: Tomatoes are juicy berries, and the edible part includes the pericarp and the placenta tissue. The fruit comes in various shapes, such as spherical, oblate, elliptical, and pear-shaped. The ripe fruit is red, pink, or yellow. The external color of the fruit is formed by the combination of the color of the fruit skin and the color of the flesh. If both the flesh and the skin are yellow, the external appearance of the fruit is orangish-yellow; if the flesh is red and the skin is colorless, the external appearance of the fruit is pink; if the flesh is red and the skin is yellow, the fruit is bright red. The red color of tomatoes is due to the presence of a large amount of lycopene in the fruit. Yellow tomatoes do not contain lycopene but only contain various carotenes.

(6)Seeds: The seeds are small, flat, and kidney-shaped, and covered by silver-gray fine hairs. The seed is reddish-black or yellow, with a kidney-shaped and flat appearance, and a thousand-seed weight of about 3 g (Fig. 2-3). The germination age of the seeds can reach 3–4 years.

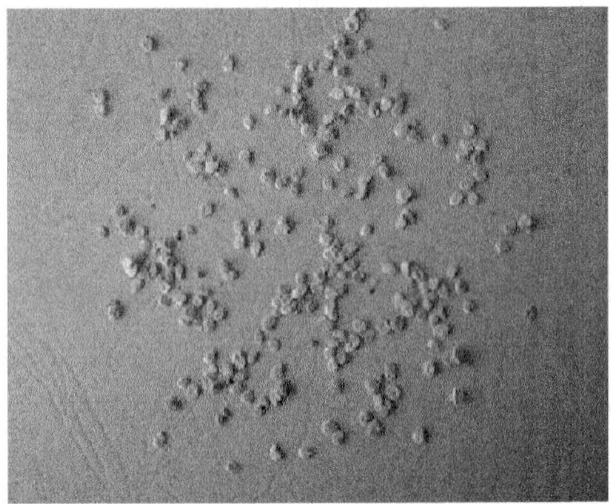

Fig. 2-3 Tomato Seeds

2. Requirements for environmental conditions

(1) Temperature: Tomatoes prefer warm temperatures and are not tolerant of heat. The ideal temperature for tomato growth is 20–25°C, which promotes the formation of lycopene in tomatoes. Different stages of growth have different

temperature requirements and responses. The optimum temperature for seed germination is 25–30°C, with a minimum germination temperature of 12°C. During the seedling stage, the suitable daytime temperature is 20–25°C, and the nighttime temperature is 10–15°C. Lower temperatures during the seedling stage, especially at night, can promote floral bud differentiation, lower the internode of the first inflorescence, and increase the number of flowers per inflorescence. We can take advantage of the strong temperature adaptability of tomato seedlings, and carry out low-temperature hardening in cultivation before transplanting seedlings to improve the quality of the seedlings. The suitable daytime temperature for flower bud differentiation is around 24°C, with a nighttime temperature of around 17°C. The optimal daytime temperature during the flowering period is 20–25°C, with a nighttime temperature of 15–20°C. Temperatures below 15°C or above 35°C are unfavorable for flower development and fruit set. During the peak fruiting stage, the desired daytime temperature is 25–28°C, with a nighttime temperature of 15–20°C. Lower temperatures result in slow fruit growth, for example, inadequate coloration below 12°C, and poor fertilization above 35°C lead to reduced fruit set and the risk of heat-induced ripening or hollow fruits. Growth ceases at temperatures above 40°C.

As for soil temperature, the most suitable range is 20–23°C, with an upper limit of 33°C. The young roots start elongating at 6°C, root hairs develop at 8°C, and rapid elongation occurs at 15–20°C, with 25–28°C as the most favorable temperature range. Increasing soil temperature within the appropriate range can promote the development of the root system. Therefore, when tomatoes are transplanted in early spring, the soil temperature must be stable above 8°C.

(2) Light: Tomatoes are light-loving plants and require sufficient sunlight for growth and development. The light saturation point is around 70 klx, and the light compensation point is around 2 klx. Insufficient light leads to excessive elongation, fewer flowers, nutrient deficiencies, flower and fruit drop, as well as more physiological disorders and diseases. Therefore, it is necessary to adjust temperature management based on the amount of sunlight. If photosynthesis is sufficient during the day, the nighttime temperature can be slightly higher. However, when

photosynthesis is inadequate during the day, the nighttime temperature should be slightly lower to avoid energy consumption.

Additionally, tomatoes do not have strict requirements for daylight duration. As long as the temperature conditions are suitable, they can be cultivated throughout the year.

(3) Water moisture: Tomato plants have a large number of leaves and a significant leaf area, resulting in strong transpiration. Additionally, since the fruit is a berry and the plant produces a large number of fruits, tomatoes require a substantial amount of water. The water requirements of tomatoes vary at different stages of growth and development. The soil humidity should be around 60% during the seedling stage and 80% during the fruiting stage, and excessive or drastic fluctuations in soil moisture during fruit ripening can lead to dehiscent fruit and reduce their commercial value. During the fruiting stage, it is crucial to maintain adequate soil moisture for tomatoes, ensuring a consistent moisture level and preventing alternating periods of dryness and wetness. Tomatoes are not tolerant of waterlogging. If the field is waterlogged for 24 hours, the roots may suffer from oxygen deficiency and die due to suffocation. Therefore, deep furrows and high ridges should be created to prevent waterlogging and ensure proper drainage after rainfall.

Tomatoes prefer a relatively dry climate, with the relative air humidity maintained at 45%–50%. Excessive humidity of air leads to weak plant growth, delays development, hinders proper pollination, and increases the occurrence of diseases, especially under high temperature and high humidity conditions. Therefore, when cultivating tomatoes in a controlled environment, we should pay special attention to ventilation and air exchange to prevent excessive humidity and serious diseases.

(4) Nutrient conditions of soil: Tomatoes have strong adaptability to soil, but they prefer deep, organic-rich, well-drained, and well-ventilated loamy soil or sandy loam soil. The optimal soil pH for tomatoes is in a range of 6.5–7.0. Tomatoes grown in sandy loam soil tend to mature earlier, while those grown in clay loam soil have higher yields. However, it is not recommended to cultivate tomatoes in clay or low-lying areas.

Tomatoes have a long growth period and require the absorption of large amounts of organic nutrients and various inorganic nutrient elements in order to achieve high-quality fruits with high yields. The ratio of nitrogen, phosphorus, and potassium in fertilization should be 1 : 1 : 2 for optimal results. Additionally, the lack of micronutrients can cause physiological diseases.

II. Types and Varieties

According to the growth habits of plants, tomatoes can be divided into two types: indeterminate growth and determinate growth (including self-topping and high-topping).

Currently cultivated tomatoes are common tomatoes with the following five variants.

(1) Cultivated tomatoes. They have large fruit sizes, ranging from oblate to spheric shape, with colors including bright red, pink, deep red, light yellow, deep yellow, etc. The plants have trailing stems and many branches. This is the most common variant for cultivation.

(2) Large-leaved tomatoes. Their leaves are larger in size, fewer in number, without lobes, and resemble potato leaves. The fruit characteristics, stems and branching are similar to those of common tomatoes.

(3) Upright tomatoes. Their stems are upright, reaching a height of 60–75 cm. The leaves are small and thick, with a dark green color, and the fruit is oblate in shape.

(4) Cherry tomatoes. They have small and round fruits, resembling cherries, with two chambers. The plants are robust, with slender stems. Their leaves are small and light green. The fruits come in yellow, red, and other colors.

(5) Pear-shaped tomatoes. The plants have small fruits shaped like pears, in colors such as red and orangish-yellow. The plants are vigorous, with dense green and small leaves.

III. Cultivation Season and Methods

Tomatoes have lenient requirements for photoperiods and can blossom and fruit throughout the year as long as the temperature is suitable. In open fields, they are typically divided into spring tomatoes and autumn tomatoes due to the separation of winter and summer, resulting in a relatively short supply period. With the improvement of people's living standards, the market demands balanced and year-round production of fresh tomatoes. In recent years, the use of techniques such as polytunnels, small- and medium-sized greenhouses, mulching, sunlight greenhouses, and shade nets in the Yangtze River basin has made the tomatoes achieve delayed autumn production, winter insulation, early spring production, and summer sun shading, effectively extending the cultivation and supply time of tomatoes. Numerous valuable experiences in segmented sowing, multi-layer covering, precise management, and high-quality and high-yield production have been created in various regions. In many areas, high-mountain cultivation and coastal cultivation of tomatoes, which are carried out by utilizing the abundant terrain and landforms in the southern regions, settle the problem of discontinuous tomato cultivation caused by high temperatures in summer, which has basically achieved balanced production and year-round supply of tomatoes in these regions. The following are examples of commonly used cultivation practices in representative areas of the Yangtze River basin, listed in Table 2-1.

Table 2-1 Cultivation Season and Cropping Schedule for Tomatoes in the Yangtze River Basin

Season and Cropping Schedule	Sowing Phase	Transplanting Phase	Harvest Phase
Spring Crop	Early December – Mid - January	Mid – March	Mid - May – Mid - July
Autumn Crop	Mid-July	Early August – Mid - August	Late October – Late November

IV. Cultivation Techniques

1. Seedling raising technology

Cultivating strong seedlings is one of the key measures for high-yield cultivation of tomatoes. The criteria for strong seedlings are as follows: thick stems, short internodes, large and thick leaves, dark green color, many and robust roots, free from pests and diseases, undamaged, plant height of 22–25 cm, stem thickness of 0.5–0.6 cm, complete cotyledons, 8–9 true leaves with large flower buds, and seedling age of 80–100 days.

(1) Sowing.

Seedling cultivation should be prepared before September. Choose higher garden soil or paddy field soil that has not been used for the cultivation of solanaceous crops for over 3 years. Crush and mix completely decomposed organic fertilizers such as pig and chicken manure and rice hull ash in a ratio of 5 ∶ 3 ∶ 2 to make "cultivation soil". In early October, build a polytunnel and level the seedbed inside the polytunnel. Cover the seedbed with 5-cm-thick nutrient soil. Before sowing, fill the bottom of the seedbed with water, thinly spread a layer of dry fine soil, and generally sow seeds per 5–8 g/m^2 on the bed surface. For greenhouse cultivation, sow in mid- to late October, while for medium and small greenhouses, sow in early November. Soak the seeds in 55°C warm water for 20 minutes; or after being fully soaked, the seeds are disinfected with an appropriate chemical solution (1% potassium permanganate or potassium dihydrogen phosphate). Then, the disinfected seeds are rinsed with clean water and covered with damp gauze to germinate at a temperature of 25–30°C. Sow when 2/3 of the seeds have sprouted, or sow directly after the seeds are rinsed clean and dried. If dry seeds are used for sowing, sow the seeds evenly after they are mixed with chlorothalonil of 1% of the seed weight.

(2) Seedbed management.

Before emergence, maintain a daytime temperature of above 25°C and a nighttime temperature of around 18°C in the greenhouse. The seedlings will emerge in 4–6 days. After emergence, remove the mulching film in time. Cover the seedlings with a small greenhouse. Maintain a daytime temperature of 20–25°C and a nighttime temperature of 10–15°C, and ventilate and cool in time when the temperature exceeds 28°C to prevent the seedlings from excessive growth. Transplant the seedlings into nutrient bowls when they have 2–3 true leaves. A nutrient bowl with a diameter of 8–10 cm is better, and the bowls should be filled with nutrient soil in advance. In the small greenhouse, maintain a daytime temperature of 20–25°C and a nighttime temperature of 10–15°C, keeping the nutrient bowls moist and fertile.

There are two points that should be noted during the cultivation of tomato seedlings: ① As the starting temperature for tomato growth is relatively low, it is prone to excessive elongation. Therefore, in addition to controlling the seedbed temperature, it is necessary to increase sunlight and ventilation and reduce humidity (60%–70%). ② Early blight and gray mold are common diseases that occur in the later stage of seedling cultivation, which can be effectively prevented and treated with Bordeaux mixture at a concentration of 0.2%–0.25%.

(3) Grafting seedling cultivation of tomatoes.

The grafting cultivation of tomatoes starts later than that of cucumbers and eggplants, with a smaller promotion area and scale. However, with the continuous expansion of the area of tomato cultivation in a controlled environment and unavoidable continuous cropping, soil-borne diseases become severe, mainly including bacterial wilt, fusarium wilt, *Verticillium* wilt, root-knot nematode disease, etc. It is an effective way to choose resistant rootstocks and cultivate grafted seedlings. In recent years, it has been developed in some areas of China. Most of the rootstocks used are imported, such as LS-89, which is mainly resistant to bacterial wilt and fusarium wilt of tomatoes; Skulum, which is mainly resistant to fusarium wilt, root rot, cucumber mosaic virus, and root-knot nematode. There are also rootstocks selected from wild tomatoes. Grafting methods include splice grafting, cleft grafting, whip and tongue grafting, approach grafting, and side grafting.

(4) Other seedling cultivation methods.

① Seedling cultivation with cuttings. Take the lateral branches below the first inflorescence of the tomato plant and remove the leaves within 3 cm from the base. Flatten the wound at the base of the cutting, and place the cutting for 5–6 hours in a clean room. Soak it in 20,000 times NAA or 10,000 times IAA for 10 minutes, and then put it in clean water. Control the room at relatively a higher temperature and air humidity. Transplant it into the soil after new roots grow out.

② Industrial seedling cultivation. In Japan and the Netherlands, modern production equipment and continuous automatic operation technology are adopted to produce tomato seedlings in a closed, controlled environment. This method has also been applied in some areas of China. Plug seedling cultivation has been promoted in many suburban areas.

2. Field management technology

(1) Land preparation, ridge tillage, and base fertilization: Land preparation and fertilization must be completed 7–15 days before transplanting. First of all, remove any crop residues, deeply till the soil, and carefully level the field. The furrows should be deep and the ridges should be high, with the ridge surface forming a turtle-back shape. The ridge surface should be level and free of large clumps of soil, so that the mulching film can tightly cover the soil surface. At the same time, base fertilization should be applied during land preparation. Tomatoes have a long growth period and require a large amount of fertilizer, especially potassium fertilizer. The principle of fertilization is to apply more nitrogen and phosphorus fertilizers in the early stage, and increase the application of potassium fertilizer and micronutrients in the middle and late stages. The ratio of the three elements should be 1 ∶ 1 ∶ 2. Generally, before tilling the land, apply 4,000–5,000 kg of well-rotted organic fertilizer and 50 kg of vegetable-specific fertilizer per mu. After applying the fertilizers, make deep furrows and high ridges that are 1.4 m wide (including the furrow), and cover them with mulching film.

(2) Transplanting: When the soil temperature at a depth of 10 cm stabilizes at 8°C and the seedlings have 7–8 leaves, choose a period when the cold fronts have ended and the warm fronts are approaching, determine the correct row and plant spacing and density based on the variety characteristics and cultivation methods, and then dig holes for transplanting. If a combination of multi-layer coverage, including mulching film, row cover, medium arched greenhouse and polytunnel, is adopted, tomatoes can be sown in early October and transplanted in mid-December. The seedlings which are cultivated in mulching film, row cover, and polytunnel can be transplanted in early to mid-February. The seedlings which are cultivated in a small greenhouse covered by polytunnel can be transplanted in late February.

The seedlings which are cultivated in a small greenhouse and covered by mulching film can be transplanted in early March. The seedlings which are cultivated in open fields can be transplanted in mid- to late March.

Plant two rows per ridge, with a transplanting density per mu of 3,500 plants for early-maturing varieties, 3,000 for medium-maturing varieties, and 2,500 for late-maturing varieties. The planting depth should not be too deep, and the bowl surface should be level with the ridge surface. After transplantation, some root fertilizer should be applied immediately. If the mulching film is covered, make sure the opening is as small as possible. When seedlings emerge from the mulch, handle them gently. Cover and lay the mulch tightly. Press the opening and each layer of the film with soil to facilitate insulation. Additionally, construct a row cover to cover the film.

(3) Temperature and light management during the early stage of transplanting: In the first week after planting, attention should be paid to the insulation of seedlings to facilitate their growth. After seedling acclimation, maintain a temperature of around 25°C during the day and above 15°C at night. When the temperature inside the greenhouse exceeds 30°C, especially during high temperatures and humidity, ventilation should be carried out promptly. In spring, it is often cloudy and rainy in southern China, and the sunlight is relatively insufficient. When the temperature is high during sunny days or at noon, it is necessary to uncover the mulch in the greenhouse, either partially or completely, to increase the plant's exposure to sunlight. When the weather is good, uncover the mulch early and cover it late; when the weather is bad, uncover it late and cover it early. If the mulch has been used for too long and the light transmission is poor, it should be replaced promptly. In the case of cloudy and rainy weather, attention should be paid to ventilation and humidity control, so as to reduce the occurrence of diseases. In the case of cold waves and frost, it is necessary to strengthen insulation to prevent tomatoes from frost damage. Then, build trellises for the tomatoes promptly and remove the row cover based on the growth of the tomato plants.

(4) Fertilizer and water management: After the seedling acclimation of tomatoes, they can be lightly and frequently fertilized with diluted manure water.

However, for vigorously growing indeterminate varieties, it is recommended to avoid excessive nitrogen fertilizer before the first cluster of fruits set in order to prevent excessive growth. Depending on the degree of nutrient deficiency in the plants, it is recommended to apply 10–15 kg/mu of specialized vegetable compound fertilizer at the rhizosphere or through mulching film after the first and third clusters of fruits have been set. During the growth process, potassium dihydrogen phosphate (1%–2%) can be applied through extraradical spray multiple times based on the seedling condition and plant performance. Tomatoes have strict requirements for soil moisture, especially during the fruiting period, with high transpiration and a big demand for water. It is important to irrigate in a timely manner to maintain soil moisture and avoid fluctuating moisture levels, which may lead to dehiscent fruit. Drip irrigation can not only meet the water requirement of plants and save water but also lower the humidity inside the greenhouse, thereby reducing the occurrence and spread of diseases.

(5) Plant pruning and regulation: In most cases, tomatoes are trained or trellised to form a three-dimensional structure that makes full use of space. Pruning is used to regulate and control the balance between vegetative and reproductive growth by properly adjusting nutrient availability. Common pruning methods include single-stem pruning, one-and-a-half stem pruning, double-stem pruning, continuous pinching pruning, and multi-stem pruning.

In some regions, additional pruning methods such as pinching and replacement pruning for two-cluster fruit and pinching and replacement pruning for three-cluster fruit are also employed.

When tomato plants grow taller than 30 cm, they generally require support for normal growth. In open-field cultivation, bamboo poles or other support materials are used to construct four-legged trellises or herringbone trellises connected by a “backbone”. In greenhouse cultivation and soilless cultivation, plastic strings are used to tie the main stem, which is suspended on horizontally tensioned wires.

Other measures in plant regulation include branch pruning, pinching, and leaf picking. Tomatoes have strong sprouting ability in lateral buds, so branch pruning involves the timely removal of all side branches except for the ones that need to be retained. Sometimes, larger leaf areas are preserved to reduce wounds before removal, and lateral shoots are removed after two fully grown functional leaves are left. Depending on the cultivation purposes and tomato variety characteristics, when the plants have reached a certain height and have produced enough fruit clusters, the terminal bud should be removed, which is known as pinching or topping-off. Timely removal of old, yellow, and diseased leaves not only reduces the waste of photosynthetic products but also improves ventilation and light exposure conditions, which is good for photosynthesis.

The plant regulation mentioned above should be carried out on sunny days and should not be done during rainy weather or when dew has not dried yet, as it may easily induce diseases.

(6) Flower and fruit retention and thinning: Reasons for flower and fruit drop include improper fertilizer and water management, poor light transmission of the mulching film, insufficient sunlight, and cultivation techniques such as drought and waterlogging. To address these issues, it is necessary to improve fertilizer and water management, replace the old greenhouse mulching films, and take measures to maintain soil moisture and improve drainage with deep furrows. Extremely low temperatures in early spring (nighttime temperatures below 15°C) and excessively high temperatures in midsummer (nighttime temperatures above 25°C) can also cause flower and fruit drop. Plant growth regulators, such as phenoxyacetic acids (2,4-D and PCPA) and naphthaleneacetic acids (β-naphthoxyacetic acid), can be used to treat flower and fruit drop caused by temperature extremes. The concentration of 2,4-D used to prevent tomato flower drop as a plant growth regulator is 10–20 mg/kg. However, 2,4-D can cause severe damage to tender leaves and buds, so it should only be used for flower soaking or coating. PCPA can be sprayed on flowers at a concentration of 25–50 mg/kg. Its concentration should be lower under high-temperature conditions and higher under low-temperature conditions. Application of plant growth regulators 1–2 days before or after flowering not only prevents flower drop, but also promotes rapid fruit growth and uniform fruit development. If the treatment is applied when the flower buds are small, the ovary will expand slowly, resulting in smaller fruits. When half of the

flowers in an inflorescence have bloomed, the entire inflorescence can be sprayed 1–2 times. However, most fruits treated with plant growth regulators will be seedless.

Deformed flowers caused by low temperatures can exacerbate fruit deformity after the application of plant growth regulators, affecting the commercial value of the fruits. Poorly developed and deformed flowers should be thinned, and when there are enough fruit clusters, a certain number of round and high-quality fruits should be retained according to the appropriate fruit-to-leaf ratio. Other fruits that are deformed, diseased, or cracked, and will affect the market value should be promptly thinned when their inferiority becomes apparent, so as to save nutrients and promote the full development of round and high-quality fruits while increasing yield and product quality.

(7) Environmental management of cultivation in a controlled environment: Early spring cultivation, early spring maturity, delayed autumn cultivation, and winter tomato cultivation all require mulching. Using multi-layer coverings such as large and medium greenhouses with small greenhouses, straw mats covered on top of small greenhouses or non-woven fabrics added in the polytunnel can effectively increase the temperature inside the greenhouse. There should be a distance of 20–30 cm between tomatoes cultivated in winter and early spring in large and medium greenhouses to provide insulation by double-layer mulch. High temperatures in summer, especially high nighttime temperatures, seriously affect the growth and development of tomatoes, but there are significant differences among varieties in their heat tolerance. Therefore, heat-tolerant varieties should be selected for summer cultivation whenever possible. To increase the light intensity inside the greenhouse, composite films with good light transmittance and strength should be used for the greenhouse covering. In suitable conditions, high-quality greenhouse films with anti-fogging, anti-drip, and anti-aging properties can be utilized. Under high temperatures and good sunlight conditions, the mulch on the small greenhouse should be removed early and covered late. After the temperature stabilizes and rises, the small and large greenhouses should be gradually removed to increase the light exposure for tomato plants. To improve soil humidity and reduce air humidity,

the surface of the ridge should be covered with mulching film. Black and blackish-gray double-sided films can also prevent weed growth. Drip irrigation can not only reduce air humidity but also save water. If the soil humidity or the groundwater level is too high, furrows can be deepened in the greenhouse or trenches can be deepened between the greenhouse compartments. During the transition from spring to summer, on sunny midday, the temperature inside the greenhouse may reach over 40°C. It is necessary to separate the side film and top film or open the greenhouse doors for ventilation and cooling in a timely manner. Greenhouses with higher roofs and those with opening and closing devices on the roof have better cooling effects. When there is a shade net installed on the roof, the cooling effect inside the greenhouse is significant. In some areas, the top film is not removed in summer, and an additional layer of shade net is added on the top, which can not only provide good shading and cooling effects, but also effectively prevent sudden attacks from natural disasters such as storms and hails.

(8) Timely prevention and control of pests and diseases: Tomato's main diseases include viral diseases, gray mold, late blight, leaf mold, early blight, bacterial wilt, canker, etc. The common pests include green peach aphids, whiteflies, etc. Varieties with resistance or tolerance to viral diseases, gray mold, late blight, leaf mold, early blight, bacterial wilt, and Fusarium wilt are selected, such as Maofen 802, Zaofeng, Xifen, Sukang, Zheza, Fengshun, Yuesheng, Xiangfanqie No.1, etc.,

(9) Harvesting: The tomato fruits can be divided into six ripening stages: mature green stage, breaker stage, turning stage, pink stage, light red stage, and red ripe stage. After the tomatoes enter the mature green stage, there is little change in fruit weight and size until maturity. Tomatoes for long-distance transportation should be harvested from the mature green stage to the breaker stage, while tomatoes for short-distance transportation should be harvested from the turning stage to the pink stage. When they reach the destination, tomatoes should reach the market-required red ripe stage. Tomatoes sold locally can be harvested at the light red stage. Timely harvesting of fruits at the appropriate ripening stage has a significant impact on the growth rate and quality of higher-grade fruits and also

affects the total yield.

3. Soilless cultivation techniques

(1) Cultivation methods. There are many methods of soilless cultivation for tomatoes. Liquid cultivation methods such as rock wool, nutrient film, and deep water culture can be used, as well as various substrate cultivation methods such as vermiculite, perlite bag cultivation, sand cultivation, rice hull ash, puffed chicken manure, etc.

(2) Seasons and cropping schedule. In the middle and lower reaches of the Yangtze River, soilless cultivation is commonly used for spring early-maturing greenhouse cultivation and delayed autumn cultivation. For spring cultivation of tomatoes, seedlings are raised from late December to early January, transplanted in late February, and harvested from May to July. For autumn-winter cultivation, seedlings are raised in early July, transplanted in early September, and harvested from late October to December. The application of multi-layer mulches such as small greenhouses can extend the harvest until January of the following year.

(3) Variety selection. Most varieties suitable for soil cultivation in greenhouses can be used for soilless cultivation, such as Zaofeng, Zaokei, and Maofen 802 from Xi'an, Zhongshu No.4 from Beijing, and Xiafen and Subao No.1 from Jiangsu. Heat-tolerant cherry tomato varieties can be cultivated in high-temperature summer. Seedlings are raised in early to mid-March, transplanted in late April, and harvested from July to September.

(4) Planting density. The planting density is 24,000 to 52,500 plants/hm^2. Early-maturing cultivars can be planted more densely.

(5) Formula, usage, and management of the nutrient solution. Japanese Yamazaki, Penningsfeld, and British Cooper formulas can all be used for the soilless cultivation of tomatoes. During the seedling stage, the nutrient solution is supplied at half of the standard concentration, and after flowering, it is supplied at the standard concentration. During the peak fruiting period, the concentration is increased to 1.5 times, and the electrical conductivity is controlled between 1.2 and 3.5 ms/cm. For nutrient film cultivation, the solution is supplied every 45 minutes for 15 minutes. During the peak fruiting period and hot season, it can be shortened

to every 30 minutes for 15–20 minutes. A continuous supply of solution can be provided at noon during the hot season to reduce rhizosphere temperature. The pH value of the nutrient solution should be regularly tested and regulated to maintain between 5.5 and 6.0. The nutrient solution for substrate cultivation is mainly supplied by drip irrigation, with the distance between the drippers determined based on plant spacing and row spacing. Iron deficiency symptoms are prone to occurring in the early growth stage of tomatoes, and calcium deficiency symptoms are prone to occurring in the early fruiting stage. Adjusting the pH value, adding nutrient solution, and foliar spraying with chelated iron and calcium chloride can be adopted for prevention and control.

(6) Field management and plant management. Soilless cultivation is conducted in a relatively closed environment, and a disinfected and sanitary system should be established to improve environmental cleanliness and reduce diseases and the use of pesticides as much as possible. Tomatoes grown in soilless cultivation are hung on plastic rope suspension trellises. Pruning, branch pruning, removal of old leaves, and pinching are also necessary for plant regulation. Plant growth regulators can be used for flower dipping, flower and fruit thinning, and regulation of vegetative and reproductive growth. Timely harvesting of ripe fruits is also an important management task.

V. Common Cultivation Issues and Their Prevention Strategies

1.Tomato flower and fruit drop

Tomatoes are prone to flowering and fruit drop, especially flower drop, under unfavorable environmental conditions and poor plant nutrition. The reasons include: nighttime temperatures below 15°C (tomatoes growing in late autumn drop fruits); high nighttime temperatures above 25°C during summer; continuous cloudy and rainy days; excessive drought; poor nutrition; excessive vegetative growth (due to continuous cloudy and rainy days); late transplanting, resulting in excessive root injury; pests and diseases.

Preventive measures: ① Strong seedling cultivation. Strengthening seedling management and improving seedling quality are the foundation for flower and fruit retention. ② Enhance flower management during the flowering period, including proper fertilization and watering, and timely plant regulation. In spring, greenhouse cultivation requires temperature rise, insulation, and light enhancement, while in summer, shading and cooling measures should be taken to prevent high temperatures and dryness. After the fruit set, timely pruning, leaf picking and pinching, and thinning of flowers and fruits should be conducted to balance their growth. ③ Hand pollination. Between 9:00 and 10:00, shaking the plants or having people walk around to vibrate the plants can promote pollen dispersal. ④ Hormone treatment. For flower dipping, apply 2,4-D (flower dipping method and identification) at a concentration of 10–20 mg/L or spray with PCPA (dichlorophenoxyacetic acid, also known as Fanqieling or Fangluosu) at a concentration of 25–50 mg/L.

2. Misshapen fruit

Misshapen fruit includes deformed fruit, wart-like fruit, blossom-end cracked fruit, nipple-end fruit, and angular fruit.

Causes: Low temperature during flower bud differentiation, vigorous plant growth, and short daylight hours. Preventive measures: Firstly, strengthen the temperature management during the seedling stage. Secondly, firtilize and water appropriately during the fruit setting period. Thirdly, use appropriate concentrations of growth regulators correctly. Avoid repeated flower dipping, and master the timing of flower dipping. Fourthly, timely remove deformed flowers or misshapen fruits.

3. Hollow fruit

Hollow fruit refers to fruits with cavities between the pericarp and the gel-like substance of the flesh. Causes: Firstly, low temperatures during flowering prevent proper fertilization and result in hollow fruit. Secondly, hollow fruits are formed due to excessively high concentrations of growth regulators or treatment with growth regulators during the flower bud stage. Hollow fruit can also be caused by insufficient light, high temperatures during the early fruiting stage, inadequate nutrition in later stages, and improper watering during the fruiting period.

4. Blossom-end rot

Blossom-end rot is characterized by black or brown discoloration and subsequent rotting of the fruit's blossom end (near the style). Causes: ① Physiological calcium deficiency. ② Irregular water supply during the growth period can also cause blossom-end rot.

「**Summary of Sub-context**」

This sub-context mainly discusses the growth and cultivation techniques of tomatoes. It is important to grasp each technical aspect and understand the key points of tomato cultivation in the Yangtze River basin.

「**Expanded Knowledge**」

The Benefits of Lycopene

Lycopene is a type of carotenoid and a powerful antioxidant. It has strong free radical scavenging abilities and significant effects in preventing and treating cancer. It also has benefits in preventing cardiovascular diseases, improving immune function, delaying aging, etc. Lycopene is often referred to as the "golden nutrient of plants" and is acclaimed as the "new favorite in the realm of 21st-century health supplements".

Sub-context 2 Production of Eggplant

Eggplant is an annual herbaceous plant in the *Solanum* genus under the Solanaceae family, known for its berry-like fruit. It is a tropical perennial plant, also known as "luosu" and "kunlun melon" in Chinese. Its chromosome number is 2n=2x=24. The young and tender berries of eggplant are edible and can be stir-fried, boiled, and fried. They can also be dried and pickled. Eggplant contains a large amount of protein, as well as minerals such as calcium, phosphorus, and iron.

It has the effect of lowering cholesterol and enhancing liver function. The pericarp of purple eggplant also contains abundant vitamin P, which has an auxiliary therapeutic effect on patients with hypertension and purpura.

Eggplant originated in the Southeast Asian tropical region, and ancient India was its earliest domestication site. Wild and closely related species of eggplant still exist in India today. Through long-term cultivation and acclimatization, its flavor has been improved and the fruits have become larger. It was introduced to Africa during the Middle Ages, reached Europe in the 13th century, became more widely cultivated in southern Europe in the 16th century, and spread to central Europe in the 17th century. It was later introduced to the Americas. In the 18th century, it was introduced to Japan from China. China has a long history of eggplant cultivation, with various types and varieties. It is generally believed that China is the second place of origin for eggplant. Ji Han, a writer of the Western Jin Dynasty, mentioned in his botanical work *Plants of the Southern Regions* that there were eggplant trees in southern China, which is the earliest recorded information about eggplant in China. Eggplant is distributed worldwide, with the most cultivation in Asia, accounting for about 74% of the world's total production. Europe ranks second, accounting for about 14%. It is cultivated throughout China, becoming one of the main vegetables in summer.

I. Biological Characteristics

1. Botanical features

In tropical regions, eggplant is a perennial shrub, while in temperate regions, it can only be cultivated as an annual herbaceous plant.

(1) Roots: The root system is well-developed, with a thick and strong taproot. The taproot of a mature plant can reach a depth of 1.3–1.7 m, while the lateral roots are relatively short. They are distributed about 5–10 cm below the ground and have well-developed horizontally growing lateral roots, which can extend horizontally for about 1.2 m. The root system of eggplant lignifies early and has poor regenerative ability, making it unsuitable for transplantation.

(2) Stems: The stems are erect, stout, and 80–110 cm tall. They can be purple, dark purple, or green, with a strong degree of lignification.

Eggplants exhibit regular branching and fruiting habits. In early-maturing cultivars, flowering occurs when there are 6–8 leaves, while in late-maturing cultivars, flowering is observed when there are 8–9 leaves. The terminal bud transforms into a floral bud, followed by the emergence of two equally vigorous lateral branches from the adjacent axillary buds, taking over the main stem and extending growth in a fork-like shape. Afterward, a flower forms on the terminal bud at every certain leaf position, and the lateral branches also branch in the same way. As a result, the fruits grown at the first, second, third, and fourth branching junctions are called Menqie, Duiqie, Simendou, and Bamianfeng, respectively. Subsequent upward branches and increasing numbers of flowering make it difficult to count the number of fruits, and these fruits are called Mantianxing. (Fig. 2-4)

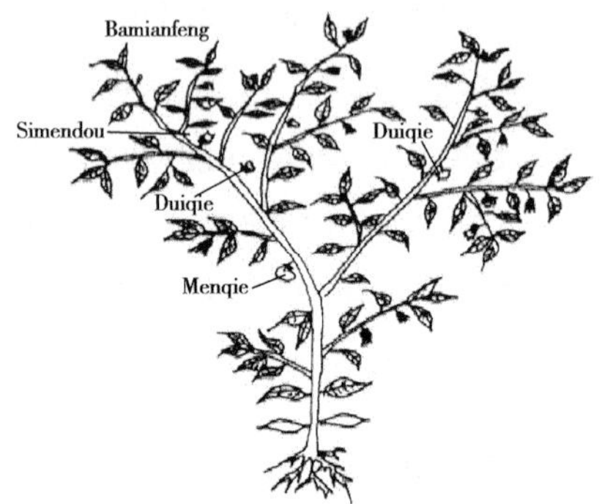

Fig. 2-4 Branching and Fruiting Habit of Eggplant

(3) Leaves: The leaves are alternate and simple, ovate or long ellipse. The leaves of tall plants are narrow and long, while those of short plants are relatively wide. The color of stems and leaves is related to the color of the fruit. In varieties with purple fruits, the tender stems and petioles have a purple color. In varieties with white or green fruits, the petioles are mostly green, but the leaf color turns green under lower temperature conditions.

(4) Flowers: The flowers are bisexual flowers. The petals and sepals are 5–6 in number, fused at the base to form a tubular shape, and the corolla is whitish-purple. There are 5 stamens, and the anthers release pollen when the flower opens (Fig. 2-5). In normal flowers, the stigma protrudes above the anthers and can be

pollinated normally. The flowers are normal when they are solitary flowers. If there are more than two clusters of flowers or if the plant is malnourished, abnormal short-styled flowers will appear, which cannot be pollinated. In mid-styled flowers, the stigma is level with the top of the anthers, and their pollination rate is lower than that of long-styled flowers. Within 2–3 days after eggplant flowers bloom, the stigma can be pollinated.

Fig. 2-5 Eggplant Flowers

(5) Fruits: The fruit consists of exocarp, mesocarp, and endocarp. The mesocarp is fleshy and pulpy, accounting for a large portion of the edible part. Eggplant fruits come in various colors such as white, green, purple, and dark purple, as well as whitish-green and whitish-purple stripes, all of which have a glossy appearance. The pigment in the fruit is anthocyanidin. The fruit matures 50–60 days after flowering, turning yellowish-brown and losing its original color and luster. Due to different varieties, the shape of the fruit also varies, including round, long, ellipse, ovate, and flattened shapes.

(6) Seeds: The seed coat has fine lines and is glossy without hairs. The aged seeds or poorly cleaned seeds during harvesting appear pale brown and lose their luster. The thousand-seed weight of seeds is 4–5 grams, and the seeds have a lifespan of 3 years (Fig. 2-6).

2. Requirements for environmental conditions

(1) Temperature: Eggplant is a warm-loving and heat-tolerant crop, with an

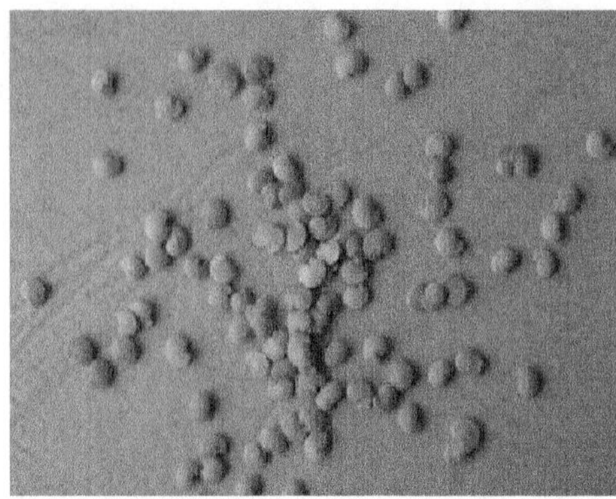

Fig. 2-6 Eggplant Seeds

optimum temperature of around 25°C during its growth and development. When the temperature drops below 20°C, eggplant plants grow slowly, which may affect their pollination, fertilization, and fruit growth. When the temperature is below 15°C, eggplant plant growth basically stops, leading to flower and fruit drop. When the temperature is below 10°C, the plant's metabolism is disrupted; below 5°C, the plants are susceptible to frost damage. When the temperature exceeds 35°C, the floral organs of eggplant will developed poorly, leading to the production of either undeveloped or dropped fruits.

Eggplants have different temperature requirements at different stages of growth and development. The suitable temperature for seed germination is 25–30°C. From emergence to the appearance of true leaves, the ideal temperature is around 20°C during the day and around 15°C at night. During the seedling stage, the optimal temperature is 22–25°C during the day and 15–18°C at night. During the fruiting period, the stem, leaf, and fruit growth require temperatures of 25–30°C during the day and 16–20°C at night.

(2) Lighting: Eggplants are not sensitive to photoperiod, meaning that the duration of daylight does not have a significant impact on their development. Eggplants have a high requirement for light intensity, with a light saturation point at 40 klx and a compensation point at 2 klx. Purple and red varieties have higher requirements for light intensity when compared to other varieties. Insufficient light intensity can affect the coloration of eggplants and their marketability. Therefore,

more attention should be paid to proper planting density in order to make full use of sunlight. Adequate light results in a glossy pericarp and vibrant color, while weak light can lead to a high flower drop rate, misshapen fruits, and darker skin color.

Furthermore, the light intensity also affects the quality of eggplant flowers. According to research, higher light intensity leads to a greater proportion of long-styled flowers, while lower light intensity results in a higher proportion of short-styled flowers. Therefore, in seedling cultivation and greenhouse cultivation of eggplants, it is recommended to provide strong light for the seedlings (plants) as much as possible.

(3) Water moisture. Eggplants have weak drought tolerance and tend to have lush foliage, large and thick leaves, vigorous transpiration, and abundant flowers and fruits. Therefore, they have high water demand. The maximum field capacity should be maintained preferably at 60%–80%, generally not lower than 55%. Otherwise, it may lead to stunted seedlings and undeveloped fruits. Eggplants are moisture-loving but are also susceptible to waterlogging. Therefore, successful eggplant cultivation requires a careful balance of irrigation during drought and drainage to control waterlogging.

(4) Soil and nutrient conditions: Eggplants do not have strict soil requirements. They generally grow best in sandy loam soil that is rich in organic matter, loose, fertile, and well-drained, and especially thrive when grown in slightly acidic to slightly alkaline (pH 6.8–7.3) soil, which results in higher yields.

Eggplants are vegetables that require a relatively large amount of nutrients. They have a long growth period, and require approximately 30 kg of nitrogen, 6.4 kg of phosphorus, 55 kg of potassium, and 44 kg of calcium to produce every 10,000 kg of marketable fruits. Calcium and magnesium are also important for the development of eggplants. Calcium deficiency can cause browning near the veins and the appearance of "rust-like" leaves. To supplement the calcium content in the soil, lime can be applied during land preparation. Magnesium deficiency in the soil can affect chlorophyll formation, resulting in yellowing near the veins, especially the main veins. Foliar application of 0.05%–0.1% magnesium sulfate solution 2–4 times can correct the magnesium deficiency.

II. Types and Varieties

Eggplants can be classified based on maturity into early-maturing, medium-maturing, and late-maturing varieties. They can also be categorized by color, such as purple, red, white, and green. Based on fruit shape, eggplants can be divided into the following three types.

(1) Round eggplants: These plants are tall with wide and thick leaves. The fruits are large, spheric, spheroidal, or ellipsoidal in shape. The skin color can be purple, dark purple, reddish-purple, or greenish-white, and each fruit weighs 0.5–1 kg. They are not tolerant of wet and hot conditions and are mainly cultivated in northern regions. Most varieties of eggplants are medium and late maturing, with main varieties including Beijing Dahongpao, Liuyeqie, Jiuyeqie, Shandong Dahongpao, Tianjin Erminqie, etc.

(2) Long eggplants: These plants have moderate growth vigor, and the fruits are long and cylindrical in shape, measuring over 30 cm in length. The skin color can be purple, green, or light green. They are tolerant of wet and hot conditions and are widely cultivated in southern regions. Most varieties of eggplants are medium- and early-maturing, with common varieties including Nanjing Zixianqie, Hangzhou Hongqie, Ningbo Tengqie, Guangdong Ziqie, etc.

(3) Oval eggplants: These plants are shorter in height, and the fruits are small and have an ovate or long ovate shape. They have more seeds and inferior quality compared to other varieties. Most of them are early-maturing varieties, such as Jinan Yiwohou, Beijing Xiaoyuanqie, etc.

Currently, the main eggplant varieties cultivated in southern China include Hangqie No.1, Hangzhou Hongqie, Yinqie No.1 and No.2, Hangfeng No.2 and No.3, Suzhou Niujiaoqie, Nanjing Zixianqie, Ningbo Tengqie, Nongyou Changqie, etc.

It should be noted that eggplant consumption and cultivation have strong regional preferences, so it is recommended to choose appropriate varieties for different regions based on local cultivation and consumption habits.

III. Cultivation Season and Methods

Eggplants have lenient requirements for photoperiods and can blossom and fruit throughout the year as long as the temperature is suitable. In China, Taiwan, Hainan, Guangdong, and the southern part of Fujian, there are long summers without low temperatures in winter, allowing year-round cultivation of eggplants in open fields. Due to the low latitude and high altitude of the Yunnan plateau, there is no hot summer, making it suitable for eggplant cultivation with a long growing season, so eggplants can be cultivated in many places throughout the winter. The Yangtze River basin has four distinct seasons, including hot midsummer and relatively cold winter. Eggplants are more heat-resistant than tomatoes and are significant vegetables for many areas to supplement the autumnal food supply due to their longer supply time in the summer. In recent years, the use of techniques such as polytunnels, small-and medium-sized greenhouses, mulching, sunlight greenhouses, and shade nets in the Yangtze River basin has made the eggplants achieve year-round production and supply.

IV. Cultivation Techniques

1. Land preparation, ridge tillage, and base fertilization

Eggplants should not be continuously planted with other solanaceous crops to avoid the spread of diseases such as seedling blight, bacterial wilt, and other soil-borne diseases. In the Yangtze River basin, common previous crops include Chinese cabbage, radish, mustard, spinach, etc., while subsequent crops are autumn and winter vegetables. In poorly drained soils, eggplant roots are prone to rotting. Therefore, after the previous crop is removed, the soil should be deeply turned over to a depth of 25–30 cm and exposed to sunlight. Then, make ridges with a width of 1.4 m (including furrows).

Eggplants vegetables require an ample amount of fertilizer. During the grain-filling stages, they require a lot of nitrogen fertilizer. Phosphorus fertilizer should be applied during the seedling stage to promote early fruiting. Additional nitrogen fertilizer can greatly increase the yield without causing excessive vegetative growth. Therefore, eggplants require sufficient base fertilizer which is mainly made of well-rotted organic fertilizer. Apply 2,000–3,000 kg of well-rotted organic fertilizer and 50 kg of special compound fertilizer for eggplants per mu as base fertilizer.

2. Seedling raising technology

Criteria for robust seedlings: The strong seedling of the eggplant has a thick stem, short internodes, 9–10 true leaves, large leaf blades, dark green color, and most visible buds.

When strong seedlings are cultivated, it is generally necessary to take insulation measures such as a small greenhouse covered by polytunnels supplemented by fermentation heating and electric heating. In South China, spring eggplants are sown from September to October, transplanted in December, and harvested from April to June. Summer eggplants are sown from February to March, transplanted from April to May, and harvested from June to August. Autumn eggplants are sown in March to April, transplanted in April to May, and harvested from July to November. Winter eggplants are sown in early August and harvested from October to December. In the middle and lower reaches of the Yangtze River basin, the appropriate sowing period for spring eggplants depends on the local climate, facility conditions, and market demand. Generally, they are sown in September to early November of the previous year, transplanted in March to April of the following year, and harvested from May to July. Eggplants with delayed autumn cultivation can be sown in mid-July and harvested from October to December.

The seeds are soaked in warm water and can germinate in about 4 days under variable temperature conditions before they are ready to emerge and be sown. Approximately 20–35 g of seeds are needed per mu. The sowing density is 13–20 g/m^2. When the seedlings have 3–4 leaves, they should be transplanted, with a density of 120–130 plants/m^2.

Key points for seedling cultivation: Prevent damping-off. Eggplant seedlings are most susceptible to damping-off when 1–2 true leaves emerge, which are mainly caused by soil-borne pathogens, excessive soil moisture, low soil temperature, soil compaction, and insufficient sunlight. Eliminating these risk factors is a proactive and highly effective measure for preventing damping-off. Root wilt is a common problem in the cultivation of eggplant seedlings. When the soil temperature is around 25°C, the root system of eggplant seedlings grows vigorously and has a strong absorption capacity. When the soil temperature drops below 12°C, the root hairs do not grow, and stop growing once the soil temperature drops below 10°C, which affects the growth of aerial parts and causes wilting. The main preventive measures should start with increasing the soil temperature to promote the normal growth of the root system, especially after transplanting, the soil temperature should be increased to facilitate root development.

Grafting of eggplants: With the continuous expansion of the area of eggplant cultivation in a controlled environment and unavoidable continuous cropping, soil-borne diseases become severe, mainly including bacterial wilt, fusarium wilt, *Verticillium* wilt, root-knot nematode disease, etc. It is an effective way to choose resistant rootstocks and cultivate grafted seedlings.

Selection of rootstocks: Red eggplant, also known as scarlet eggplant, is an early and widely used rootstock for eggplants, primarity resistant to fusarium wilt. It has a high grafting survival rate and strong stress tolerance of grafted seedlings. *Solanum torvum* rootstock is resistant to four soil-borne diseases: bacterial wilt, fusarium wilt, *Verticillium* wilt, and root-knot nematode disease. The seeds are very small and difficult to germinate, and should be sown 25–30 days earlier than the scion seedlings. CRP rootstock is resistant to four soil-borne diseases: bacterial wilt, fusarium wilt, *Verticillium* wilt and root-knot nematode disease, tolerant to waterlogging and suitable for cultivation in a controlled environment.

Grafting methods: The appropriate sowing period is determined based on the respective growth rates of rootstocks and scion varieties. Eggplants cultivated in open fields and arched greenhouses should be grafted in mid- and late March, and transplanted in mid-April; Autumnal eggplants grown in the polytunnels should be grafted in early August, and transplanted in mid-September; Eggplants cultivated in winter and spring should be grafted at the end of October, and transplanted in mid-December. The most common grafting methods for eggplants are cleft grafting and side grafting.

Cleft grafting is performed when the rootstock has grown 6–8 true leaves and the scion has grown 5–7 true leaves. The rootstock is placed on the grafting bench, leaving two true leaves. The portion of the rootstock above the true leaves is cut flat with a knife, and a vertical incision is made in the stem about 1.2 cm deep. The scion seedling is then removed, leaving 2–3 true leaves. The lower part of the scion is trimmed and shaped into a wedge that matches the cut of the rootstock. It is immediately inserted into the incision of the rootstock, aligned carefully, and secured with a specialized grafting clip (Fig. 2-7).

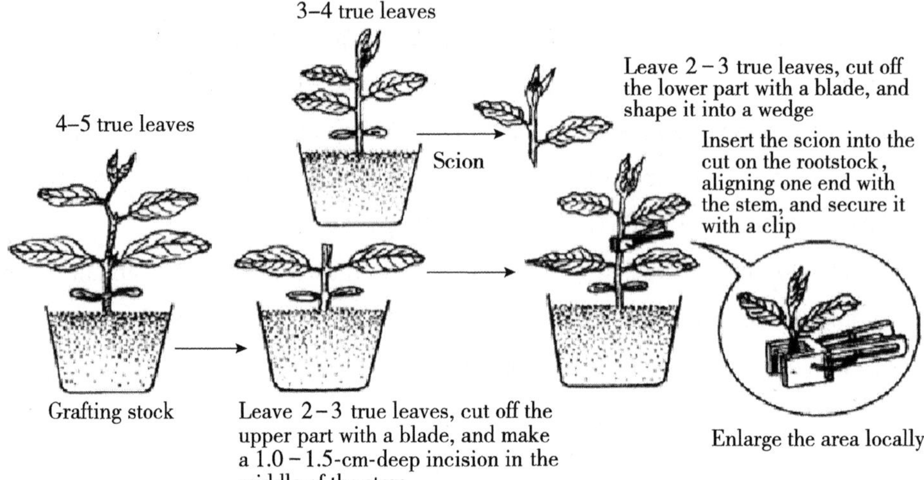

Fig. 2-7 Cleft Grafting Method for Eggplants

Side grafting follows the same age requirement as cleft grafting. During grafting, the rootstock retains two true leaves, and the upper portion is cut slantingly to form a 1.0–1.5-cm-long slant surface at an angle of approximately 30°. The scion is removed, retaining 2–3 true leaves, and the lower portion is cut slantingly to form a flat surface that matches the area and shape of the rootstock but in the opposite direction. The slant surfaces of the rootstock and scion are aligned tightly and secured with a specialized grafting clip (Fig. 2-8).

Management of grafted seedlings: To promote the survival and growth of grafted seedlings, the temperature should be controlled at a favorable range for healing, with a daytime temperature of 25–26°C and a nighttime temperature of 20–22°C. The humidity should be maintained at 95%, while the high temperatures

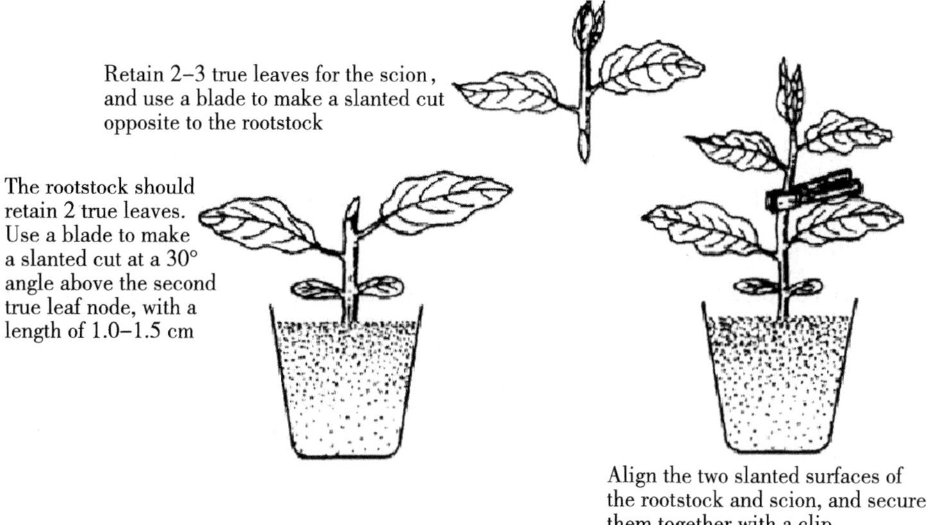

Fig. 2-8 Side Grafting for Eggplant

and direct sunlight should be avoided. The graft union can heal within 9–10 days. Avoid removing the clips too early, and loosen the bindings of healed grafts tied with plastic strips to avoid any impact on growth. Other management measures are generally similar to those for regular seedling.

3. Field management

(1) Transplanting. In the middle and lower reaches of the Yangtze River, seedlings cultivated in big, medium, or small row covers are transplanted from January to February; Seedlings cultivated in a small arched greenhouse covered by polytunnels are transplanted in the early and mid-March; Seedlings cultivated in a small greenhouse covered by a medium-sized greenhouse are transplanted in mid-March; Seedlings cultivated in a small greenhouses with mulching films are transplanted in late March; Seedlings in open fields are transplanted in the early and mid-April. The land should be prepared 10 days before transplanting, and covered by the greenhouse for preheating. Based on soil fertility, the organic fertilizer is primarily applied together with compound fertilizer to provide sufficient base fertilizer, and furrows and ridges are also made for planting. Choose a period when the cold fronts have ended and the warm fronts are approaching, determine the proper row and plant spacing and density

based on the variety of characteristics and cultivation methods, and then dig holes for transplanting. After transplantation, it is necessary to water the roots properly for their development. If the mulching film is covered, make sure the opening is as small as possible. When seedlings emerge from the mulch, handle them gently. Cover and lay the mulch tightly. Cover the small greenhouse. Press the opening and each layer of the film with soil to facilitate insulation.

(2) Temperature and sunlight management. Eggplants are not cold-tolerant. In the first week after planting, we should pay attention to the insulation of seedlings to facilitate their growth. After seedling acclimation, maintain a temperature of around 28°C during the day to promote new root growth and above 15°C at night. When the temperature inside the greenhouse exceeds 30°C, especially during high temperatures and humidity, ventilation should be carried out promptly. In spring, it is often cloudy and rainy in South China, and the sunlight is relatively insufficient. When the temperature is high during sunny days or at noon, it is necessary to uncover the straw mat and small greenhouse, either partially or completely, to increase the plant's exposure to sunlight. When the weather is good, uncover the mulch early and cover it late; When the weather is bad, uncover it late and cover it early.

(3) Fertilizer and water management. Eggplants have dense foliage, a long growth period, a high fruit set, and a high yield. Harvesting of tender fruits requires an adequate supply of nitrogen. During the seedling stage, phosphorus and potassium should be applied multiple times to promote early fruiting. Eggplants prefer fertile soil and can tolerate a high level of fertilization. They require regular application of nitrogen fertilizer, but excessive growth is rarely observed. Therefore, before transplanting, based on the original fertility of the field, the fertilizer application and nutrient consumption of previous crops, the organic fertilizer is primarily applied, together with compound fertilizer containing phosphorus and potassium, to provide sufficient base fertilizer. During the peak fruiting stage, topdressing can be applied based on fruiting and plant nutrient deficiency, combined with intertillage and soil cultivation. Apply 10–15 kg of microbial compound fertilizer per mu field per time. Eggplant has large leaves, with high transpiration and a big demand for water during the fruiting period, so it is

important to irrigate in a timely manner to maintain soil moisture. In southern regions with abundant rainfall, eggplants are not tolerant of waterlogging, so deep furrows and high ridges should be applied for better drainage. In areas prone to downpours and with lower terrain, drainage machinery should be equipped.

(4) Plant regulation, flower drop and its prevention. Pruning and leaf picking can improve ventilation and light penetration, optimize the planting structure in the field, save nutrients, reduce diseases, increase flower bud formation, enhance fruit setting, and improve quality. After the fruit setting of "Duiqie" (the first fruit), remove the lateral branches below "Menqie" (the second fruit). When "Menqie" grows to about 4–5 cm in size, remove the old leaves below it. When "Simiandou" (the third fruit) is about 4–5 cm in size, remove the old, yellow, and diseased leaves, excessively dense leaves, and slender branches below "Menqie". The removed branches and leaves should be burned in a centralized manner. In some areas, the main stem is pruned again around the beginning of autumn to promote the growth of new shoots, so as to utilize the residual stump for regrowth in delayed autumn cultivation.

Low temperatures, low sunlight, soil drought, and nutrient deficiency can all cause flower drop. Structural defects in the floral organs are a special problem in eggplant production. Short-styled flowers prevent their pollen from reaching the stigma for fertilization, resulting in flower drop. Temperatures below 17.5°C or above 40°C can cause pollen tube growth to cease, leading to flower drop. To prevent flower and fruit drop, plant growth regulators can be applied to the flower for fruit retention. Additionally, targeted field management can be implemented based on the causes, for the purposes of improving fertilizer and water supply conditions, optimizing ventilation and light penetration, and adjusting microclimate conditions. During the low-temperature stage, preserving flowers and fruits becomes a key measure to ensure the eggplant yield. 2,4-D, PCPA, and their formulations, which are applied according to their instructions, can prevent flower drop caused by low temperatures and light intensity in the early stage.

(5) Prevention and control of pests and diseases. Common pests and diseases in eggplant cultivation include gray mold, phytophthora rot, sclerotinia rot, bacterial wilt, fusarium wilt, aphids, yellow tea mites, red spider mites, and thrips, which

should be promptly controlled.

(6) Harvesting. Eggplants are vegetables that are harvested multiple times for their tender fruits. Timely harvesting of mature fruits for commerce is crucial for increasing the yield and quality. For purple and red varieties, the width of the white margin along the persistent sepal that has not yet developed anthocyanidin can be used to determine the fruit maturity. A wider white margin on the fruit indicates rapid growth, with insufficient time for anthocyanin to develop, so the fruit remains tender. When there is no white margin along persistent sepals of the fruit, it indicates that the fruit has aged and its edible value has diminished. The best time to harvest fruits is in the morning, followed by evening. Do not harvest when the temperature is high at noon so as to extend its shelf life of the market.

V. Common Cultivation Issues and Their Prevention Strategies

Early spring eggplants are easily affected by low temperatures and other environmental conditions, which may lead to flower drop or misshapen fruits, and seriously affect yield and quality.

1. Causes of flower drop or misshapen fruits

(1) Environmental conditions. Long-term weak light in early spring or high nighttime temperatures during the seedling stage can lead to short-styled flowers. Soil drought and dry air hinder flower development, and high humidity and prolonged exposure to high humidity can affect pollination, all of which will cause flower drop.

(2) Nutritional factors. Insufficient nutrition can result in weak growth vigor, small flowers, and short style, which makes the flowers prone to dropping. Excessive nutrition can also lead to flower drop due to excessive plant growth.

(3) Hormone treatment. Late treatment, high concentration, and high temperature during treatment can lead to the formation of misshapen fruits.

2. Measures to prevent flower drop or misshapen fruits

(1) Improve the environmental conditions. Keep the mulch clean to increase the transmittance, and the straw mat is uncovered in the morning and covered in the evening to extend the light exposure time as soon as possible. Cover the ground with mulching film to increase ground-level light exposure, and provide artificial

lighting. Proper watering can keep the soil and air moist, and ventilation after watering can reduce the humidity in the greenhouse.

(2) Strengthen the water and fertilizer management to ensure adequate nutrient supply, which can promote the plant to grow vigorously without excessive growth.

(3) Hormone treatment should be carried out on the day of flowering or 1–2 days in advance. The concentration of PCPA (anti-falling drug) should be 40–50 mg/L, with high concentration under low temperature and low concentration under high temperature.

「**Summary of Sub-context**」

This sub-context mainly focuses on the requirements of eggplants for environmental conditions and key points of cultivation techniques.

「**Expanded Knowledge**」

Health Benefits of Eggplants

(1) Prevent the capillary rupture. Eggplants contain a large amount of vitamin P, which can enhance immunity, prevent capillary bleeding, and maintain normal cardiovascular function.

(2) Prevent and treat gastric cancer. Solanine in eggplants can inhibit the proliferation of tumors in the digestive system, and has a positive effect on preventing and treating gastric cancer.

(3) Remove freckles and prevent aging. Vitamin E in eggplants can prevent bleeding and aging. Regular consumption of eggplants can prevent an increase in cholesterol levels in the blood, and has positive implications for slowing down the aging process in the human body.

(4) Regulate blood pressure. Eggplants contain a large amount of potassium, which can regulate blood pressure and heart function, and prevent heart disease and stroke.

Sub-context 3 Production of Chili Pepper

Chili pepper has many aliases, such as *Capsicum annuum*, green pepper, Sichuan pepper, chili, chili eggplant, etc. It originated in Central and South America, including Mexico and Peru, and is now widely cultivated around the world. It was introduced to China through the Silk Road in the late Ming Dynasty, and is now commonly grown throughout China. Chili is cultivated extensively in southern China, especially in Jiangxi, Hunan, Sichuan, Jiangsu, Zhejiang, and other regions. With high nutrition, chili has the highest vitamin C content among vegetables. It is considered a healthy food, for its roots, stems, seeds, and fruits can all be used for medicinal purposes. Chili pepper is also one of China's major export products, and enjoys a good reputation internationally. It can be dried, pickled, and made into sauce.

I. Biological Characteristics

1. Botanical features

(1) Roots: The taproot of the chili plant is underdeveloped, with fewer roots and shallow soil penetration. The root system is mainly distributed within the top 30 cm of the plow layer. Its regenerative ability is weaker than that of tomatoes and eggplants.

(2) Stems: The stem is upright, yellowish-green in color, with dark green longitudinal stripes. Sometimes the stem is purple. The base of the stem lignifies and is relatively tough. The stem is upright, and its base often lignifies. Generally, it branches into pseudo-dichotomous or trichotomous patterns. Based on branching habits, chili peppers can be divided into two types: indeterminate branching and determinate branching. Except for cluster peppers, the majority of cultivars belong to the indeterminate branching.

(3) Leaves: The leaves are alternate, and ovate, lanceolate, or elliptical in shape, with pointed tips. The leaf surface is smooth and slightly shiny. Bell peppers have longer and wider leaves, which are more scattered and erect on the plant.

(4) Flowers: The flowers are solitary or clustered. The corolla is white or greenish-white, radially symmetrical. The calyx fused at the base forms a bell-shaped calyx tube, with 5 persistent teeth at the apex. The anthers are purple, and the style is longer than the stamens. Chili peppers are easily cross-pollinated and have a natural hybridization rate of 10% through insect pollination.

(5) Fruits: The pericarp often separates from the placental tissue, forming relatively large cavities. Long-fruited varieties generally have 2 chambers, while round chili peppers and bell peppers have 2–4 chambers. When immature, chili fruits are green in color, and they turn red, yellow, brown, or purple when they are ripe. The fruits of some varieties can exhibit various colors such as green, yellow, and red on the same plant due to different degrees of ripeness, such as the five-color pepper. The fruit can be cone-shaped, short-cone-shaped, horn-shaped, elongated, cylindrical, or prismatic. The fruit apex can be pointed, blunt-pointed, or blunt.

(6) Seeds: The seeds are flat, kidney-shaped, pale yellow, and shiny on the surface. They weigh around 6–7 g per thousand grains, with seed lifespan of 2–3 years (Fig. 2-9). It is recommended to use seeds that are 1–2 years old for production.

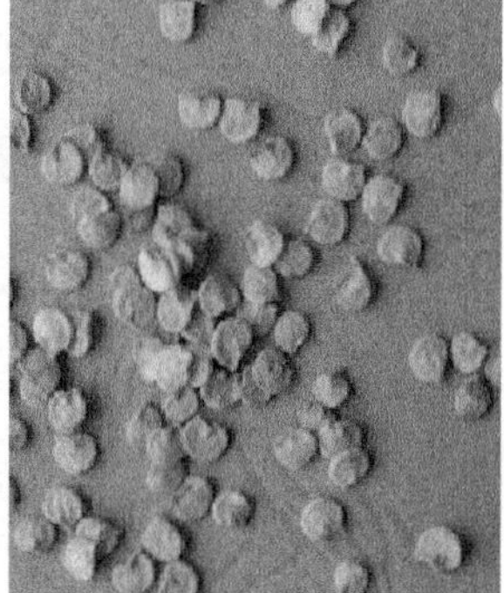

Fig. 2-9 Leaves, Flowers, and Seeds of Chili Pepper

2. Requirements for environmental conditions

(1) Temperature: Chili peppers are warm-loving vegetables. They thrive in warm temperatures and are sensitive to cold, but they are not well-suited for

extremely hot climates. Chili peppers exhibit a stronger tolerance to temperature variations compared to tomatoes and eggplants. Their critical temperature range for growth is 12–35°C, with an optimal range of 20–30°C. The ideal temperature is around 22–27°C during the day and 15–20°C at night. Different growth stages require different temperatures. The optimal temperature for seed germination is 25–30°C. After developing 3–4 true leaves, chili plants can tolerate temperatures above 0°C without frost damage. However, higher temperatures during the seedling stage are beneficial for vegetative growth. The suitable temperature range for seedling growth is around 25–30°C during the day and 15–18°C at night. During the early flowering stage, the optimal temperature is 20–25°C during the day and 15–20°C at night. During the flowering and fruiting period, temperatures below 10°C make it difficult for pollination and fertilization, which can result in flower and fruit drop. When temperatures exceed 30°C, incomplete development of the floral organ or drying of the stigma can cause failure in pollination, leading to flower drop. For fruit development and coloration, the suitable temperature range is 20–30°C, with a requirement for day-night temperature differences. It is best to have temperatures around 26–30°C during the day and 16–20°C at night. Different varieties have different temperature requirements. Generally, large-fruited varieties are less heat-tolerant than small-fruited varieties. Therefore, chili peppers can successfully overwinter in southern China. In the middle and lower reaches of the Yangtze River basin, chili peppers can continue flowering and fruiting until the first frost in autumn and afterwards.

(2) Light: Chili peppers have light intensity requirements lower than tomatoes and eggplants. The light saturation point is around 35–50 klx, and the light compensation point is 1,500–2,000 lx. Within the suitable range, increasing light intensity can improve the fruit set rate and quality. However, light intensity above 60 klx, combined with high temperatures, can lead to viral diseases. Chili peppers are not strict in their requirements for daylight length. They can flower regardless of whether the daylight is long or short in natural conditions. Therefore, as long as the temperature is suitable, chili peppers can be grown year-round. However, poor lighting conditions remain a significant cause of flower drop. Supplemental lighting

for 16–20 hours can significantly increase early yields in indoor cultivation of bell peppers during spring. However, continuous lighting for 24 hours will reduce yields.

(3) Water moisture: Chili peppers have shallow and weak root system, so they are not resistant to either drought or waterlogging, and have strict requirements for water. Waterlogging or drought in the soil is not conducive to the growth of chili peppers. Generally, large-fruited varieties have stricter water requirements than small-fruited ones. The water demand of chili peppers varies at different stages of growth. During the seedling stage, plants do not require much water. Excessive soil humidity will make the root system develop poorly, leading to excessive plant growth. During the early flowering stage, plants grow fast and their water demand increases accordingly, but excessive humidity can cause flower drop. During the fruit expansion stage, sufficient water is needed. Insufficient water supply often causes wrinkles, curves, and dull color on the fruit surface, as well as slow expansion. Chili peppers do not have strict requirements for soil and fertility.

Air humidity also has a significant impact on the growth and development of chili peppers. During the flowering period, excessive humidity or dryness can affect normal pollination and fertilization, leading to flower and fruit drop. High air humidity in greenhouse cultivation can easily cause various diseases. The suitable air humidity for the growth and development of chili peppers is 60%–80%.

(4) Soil and nutrient conditions: The root system of chili peppers is weak and has poor absorption ability, which makes them suitable for loose, fertile, water-retaining, and nutrient-rich neutral to slightly acidic soil. Chili peppers require moderate amounts of fertilizer with moderate absorption ability, but they have strong nutrient tolerance. Adequate nutrition is the guarantee of its high yields. Chili peppers require a proportion of nitrogen, phosphorus, and potassium in a ratio of 1 ∶ 0.5 ∶ 1. According to reports, for every 5,000 kg of chili peppers produced, approximately 26.5 kg of nitrogen, 7 kg of phosphorus, and 35 kg of potassium are required. Chili peppers are also sensitive to elements such as calcium, magnesium and boron. The suitable soil pH value for chili peppers is 6.1–7.6. Bell peppers have

higher requirements for soil and water conditions than chili peppers.

In addition, the spicy taste of chili peppers is easily affected by nitrogen fertilizers. Increasing nitrogen fertilization can reduce the spicy taste.

(5) Gas conditions: Oxygen has a significant impact on the germination of chili pepper seeds. If the seedbed soil is compacted or has too much water, it is difficult for chili pepper seeds to germinate when the oxygen content in the soil is less than 10%. In greenhouse cultivation, carbon dioxide fertilization can significantly increase the yield of chili peppers. Harmful gases such as SO_2, Cl_2, and C_2H_4 in industrial waste gas and NH_3 produced by improper fertilization in greenhouse cultivation will have toxic effects on chili peppers.

II. Types and Varieties

Chili peppers are divided into sweet peppers and dried peppers according to their production purposes. According to fruit shape, they can be divided into five categories: cherry pepper, cone-shaped pepper, longhorn pepper, cluster pepper, and lantern pepper. At present, the most commonly cultivated varieties are lantern pepper and longhorn pepper. According to the spiciness of the fruit, they can be divided into three categories: bell pepper type, semi (mildly) spicy type, and pungent type. According to the maturity period, chili peppers can be classified into early-maturing, medium-maturing, and late-maturing varieties.

1. Bell pepper type

Belonging to the lantern pepper species, the plants are robust and tall, with thick and ovate leaves. They have large and white flowers and short and thick fruit stems. The big fruits can be flat-round, ellipse, persimmon-shaped, or lantern-shaped, with a concave apex. The pericarp is dark green with 3–4 longitudinal grooves, and turns red or yellow when fully ripe. The flesh is thick and sweet, suitable for stir-frying as fresh vegetables. They are cultivated in North China, East China, Northeast China, and other regions.

2. Semi (mildly) spicy type

Most of them belong to the longhorn pepper or lantern pepper species. The plants are medium-sized and slightly spreading, with the fruits hanging down.

They are cone-shaped to longhorn-shaped, with a concave or pointed tip. As they have thick flesh, and a spicy or mildly spicy taste, they can be used for stir-frying, pickling, or making sauces, suitable for most people's tastes. These varieties are mainly distributed in the middle and lower reaches of the Yangtze River basin.

3. Pungent type

The plants are shorter with many branches, and narrow and long leaves. The fruits grow upwards or obliquely clustered, and they are slender and either horn-shaped or cone-shaped, with a pointed tip. The pericarp is thin, with many seeds. The immature fruits are green, while the mature ones turn red or yellow, and they have a strong spicy taste. They can be processed into chili powder or dried chili, and are widely distributed, with the largest cultivation area in Southwest China and Central South China. There is a wide variety of cultivars.

III. Cultivation Season and Methods

Chili peppers and bell peppers can be grown in open fields in the middle and lower reaches of the Yangtze River basin for two seasons: spring and summer. For spring planting, the seeds are sown in early and mid-November of the previous year. The seedlings are transplanted in early to mid-April, and the fruits are available for market from late May to late October. For summer planting, the seeds are sown in early March, and the seedlings are transplanted in mid- to late May, and the fruits are available for market from late June to late October (Table 2-2). The multi-layer insulation measures, including mulching films, small greenhouses, and large greenhouses, can shorten the seedling period, allow for early transplanting, and adjust the market availability. In the northern regions of Hubei, Anhui, and Jiangsu, where rapid development of sunlight greenhouses has occurred, seedlings are grown in mid- to late August and transplanted in early to mid-October, with availability from January to March. Chili peppers can also be grown in winter with multi-layer greenhouse insulation. Seedlings are grown in late July and transplanted in September, with availability from January to March. Similar to eggplants, chili peppers can be pruned in July and August for regeneration, and harvested in November, which allows for year-round production and supply of fresh peppers and bell peppers in most southern regions.

Table 2-2 Cultivation Season and Cropping Schedule for Chili Peppers in the Yangtze River Basin

Season and Cropping Schedule	Sowing Phase	Transplanting Phase	Harvest Phase
Spring Planting	Early to mid-November	Early to mid-April	Late May to late October
Summer Planting	Early March	Mid- to late May	Late June to late October

IV. Cultivation Techniques

1. Soil selection

The soil suitable for pepper production should be well-drained light loamy soil or sandy loam soil that has not been planted with solanaceous crops for 2–3 years with good water retention, fertility, high oxygen supply capacity, loose texture, and high organic matter content. At the same time, the soil should be hygienic and free from pests, diseases, and harmful substances.

2. Seedling raising

During the seedling period of chili peppers and bell peppers, their growth is slower than that of tomatoes, so they are not prone to excessive growth, and they require less fertilizer and water. From sowing, emergence, to transplanting, the seedling period is longer in autumn and winter for the seedbeds are cold, and shorter in summer and when electric hotbeds are applied. However, the strong and healthy seedlings that meet the standards should be cultivated.

The standards for strong pepper seedlings are as follows: The seedling is 15 cm tall with 8–10 robust, elongated, green, and glossy leaves. The internodes are short and the rhizome measures over 6 mm. The root system is well-developed, showing large buds and a small amount of flowering.

The seeds are disinfected by soaking in warm water or dry heat treatment. Whether timely transplanting is possible when the seedlings are of suitable age should be considered to determine the sowing period for chili peppers. The key to timely transplanting lies in the cultivation facilities and insulation management measures. Generally, seeds are sown in the period from late October to early November. After disinfection, the seeds are germinated under suitable temperature conditions for about 4 days. Once most of the seeds have sprouted, they can be sown. The sowing quantity is about 60–80 g per mu, with a sowing density of 15–25 g/m^2. When the seedlings have 3–4 true leaves, they can be transplanted. Since pepper branches and leaves are fragile and easily break, special care should be taken in the operation and management during the seedling period to avoid damaging them and affecting the final seedling survival rate.

3. Proper planting density and increasing fruit setting positions

The plants of chili peppers and bell peppers are shorter with smaller leaves compared to tomatoes and eggplants, so they should be densely planted for higher yield. In terms of yield composition, bell peppers, and large-fruited peppers are weight-based peppers, so the yield cannot be increased until each fruit grows fully. In contrast, small-fruited varieties, which are number-based peppers, rely on a higher number of fruits to increase the yield. The fruits of chili peppers and bell peppers grow at branching points. So the more the branching, the more the fruit setting positions, which may result in higher yield. The number of fruit setting positions can be increased if the following measures are taken, including enhancing the early-stage nutrient and water management, promoting the budding, removing unnecessary lateral buds, reducing nutrient consumption, and promoting effective branching.

When the soil temperature at a depth of 5 cm stabilizes at around 13–15°C, the seedlings can be transplanted. In addition, during the suitable planting period, it is better to have more than 3 consecutive sunny days, and we should make efforts to take the opportunity of good weather when the cold fronts have ended and the warm fronts are approaching for timely transplanting. Transplanting is not allowed on days with strong winds or heavy rain. Generally, seedlings are planted on 2 rows on a ridge of 1.4 m wide (including the furrow), with a spacing of 25–30 cm between plants and 1 plant per hole. It is important that the seedlings be planted shallowly, with the rhizome level at or slightly higher than the ridge surface. After transplantation, it is recommended to apply some root fertilizer, cover the seedlings with mulching film, and then cover them with a small greenhouse film to facilitate insulation and promote new root growth and early seedling acclimation.

4. Field management

(1) Soil, fertilizer, and water management. Before transplanting, sufficient organic fertilizer should be applied as the main base fertilizer, with a certain proportion of phosphorus and potassium fertilizer. Both chili peppers and bell peppers are usually harvested multiple times, so additional topdressings are also necessary in addition to the base fertilizer. A common practice is to apply a light seedling-boosting fertilizer once during the seedling stage, but excessive nitrogen fertilizer should be avoided to prevent leaf and flower drop caused by excessive vegetative growth and inhibition of reproductive growth. During the fruiting stage, the frequency and quantity of topdressing should be increased to ensure continuous plant growth and fruit expansion. Generally, each harvest should be followed by one additional topdressing. Fertilizer can be applied in holes or along furrows. After fertilization, ventilation should be strengthened to avoid ammonia toxicity. Fertigation carried out in a subsurface drip irrigation system will have a better effect.

In terms of water management, chili peppers have relatively weak root systems, so frequent irrigation or drip irrigation is necessary during hot and drought seasons. Excessive rainfall should be drained promptly. After the seedlings have been acclimated, water should be controlled appropriately to promote deep rooting and slow the growth of aerial parts, which can harden the seedlings and result in sturdy plants. During the early flowering and fruit setting stage, only moderate watering is needed to balance vegetative and reproductive growth and improve the early fruit setting rate. After a large number of fruits have been set, sufficient water supply is necessary, for water deficiency will cause the pericarp to wrinkle and bend, with a dull color, which affects both yield and quality. In general, the relative soil humidity should be maintained at around 80% to meet the needs of fruit development.

(2) Flower and fruit drop. Flower and fruit drop are common issues in chili

pepper production in the middle and lower reaches of the Yangtze River. In early spring, low temperatures, insufficient light in the greenhouse, and excessive planting density may affect photosynthesis. High temperatures and drought in summer may lead to insufficient water supply. Excessive nitrogen fertilizer may cause excessive growth. Poor development or defects in floral organs, pistils, and ovules can cause flower and fruit drop. To address the above reasons, appropriate cultivation measures should be taken, including selecting varieties that are tolerant to low temperatures and weak light, reasonable planting density, scientific fertilization, enhanced water management, timely prevention and control of pests and diseases, and the application of plant growth regulators for prevention. Leaf drop may also occur due to various diseases in addition to the above reasons. Bell peppers have large fruits and sparse leaves. After rain, sunlight can cause "sunburn" when focused through water droplets. Therefore, it is advisable to intercrop bell peppers with tall-stem crops to reduce direct sunlight.

(3) Environmental management in greenhouse cultivation. Similar to tomatoes, chili peppers require the use of greenhouse film for early spring cultivation, early spring maturity, delayed autumn cultivation, and overwintering. There should be enough distance between chili peppers cultivated in winter and early spring in large and medium greenhouses to provide insulation with double-layer mulch. Multi-purpose composite films with good light transmittance and strength should be used for the greenhouse covering to increase the light intensity inside the greenhouse. Under high temperatures and good sunlight conditions, the mulch on the small greenhouse should be removed early and covered late, while in low-temperature and cloudy weather, it should be covered early and removed late. After the temperature stabilizes and rises, the small and large greenhouses should be gradually removed to increase the light exposure for plants. The ridge surface should be covered with mulching film to reduce the water evaporation from the soil, effectively increase the soil humidity, and lower the air humidity. If the soil humidity or the groundwater level is too high, trenches can be deepened between the greenhouse compartments, or separating furrows can be deepened between the greenhouses to lower the groundwater level in the planting ridges.

For chili peppers growing in a greenhouse in autumn and winter, their seedlings should be cultivated during the hot season (July and August). Shading nets and double-layer film covering are commonly used to achieve the purposes of shading, cooling, moisture retention, and protection against torrential rain. High-ridge cultivation and proper control of water and fertilizer are important measures for disease prevention and high yields in autumn-winter chili pepper cultivation in the greenhouse. In autumn, as the temperature decreases, the large greenhouse should be closed first, followed by the small greenhouse. A straw mat should be added on the small greenhouse, and another layer of agricultural film should be added on the straw mat.

(4) Plant regulation. The chili pepper generally grows vigorously. To prevent lodging, a small bamboo stick can be inserted next to each plant to support it. To improve ventilation and light conditions and maintain the ideal individual and group structure, chili pepper plants need to be regulated using the following methods: leaf removal, pinching (topping-off), and pruning. Leaf removal mainly involves removing some diseased, damaged, and old leaves from the bottom. Pruning involves cutting off crowded and overlapping branches. Topping-off is an effective method adopted in the later stage of growth to ensure a concentrated nutrient supply to the fruits. Plant regulation should be carried out on sunny days, which is beneficial for wound healing and reduces the occurrence and damage of pests and diseases.

(5) Prevention and control of pests and diseases. The main pests and diseases in chili pepper cultivation include damping-off, seedling blight, viral diseases, bacterial wilt, blight, anthracnose, as well as aphids, yellow tea mites, and thrips. Timely prevention and control measures should be taken.

(6) Harvesting. Both chili peppers and bell peppers are vegetables that can be harvested multiple times. Timely harvesting of fully grown fruits with the commercial value from the lower part of the plant is beneficial for the growth of young fruits and flowering and fruit setting at the upper level. The basic criteria for harvesting green peppers are when the pericarp is light green and has a slight gloss. Chili peppers are usually harvested for the first time around 30 days after transplantation. After the initial harvest, they can be harvested every 3–5 days. In

chili pepper cultivation for the purpose of dried pepper production, chili peppers should be fully ripe before harvesting. Generally, it takes about 25 days from the flower falling to the mature green stage, and about 20 days from the mature green stage to the red ripe stage. Fully ripe chili peppers are harvested multiple times, and the yield of drying in batches is higher than that of one-time harvesting.

5. Regeneration cultivation of chili peppers

Chili peppers can be regenerated by keeping 3–4 strong main branches from selected old plants. These branches are pruned before the buds are fully developed, which promotes the sprouting of new fruit branches and the reformation of a lush framework for fruit setting. In cases where seedlings or subsequent crops are not satisfactory, this cultivation method can sometimes achieve the yield equivalent to planting new seedlings.

V. Common Cultivation Issues and Their Prevention Strategies

Common problems in chili pepper cultivation are flower drop, fruit drop, and leaf drop (collectively known as "three drops"), which affect the yield of chili peppers. The main causes of this phenomenon: ① Excessive high and low temperature, which is the main cause of flower drop. Flower drop in early spring is caused by low temperatures, rainy weather, insufficient light, etc. ② Improper cultivation management measures, which often result in flower and fruit drop, such as excessive application of nitrogen fertilizer leading to excessive plant growth, overcrowded planting causing poor ventilation and light transmission, and deficiencies in nitrogen and phosphorus nutrients. ③ Unfavorable cultivation environment. High temperatures and drought in July and August, or sudden thunderstorms after dry and hot weather, can lead to imbalanced soil moisture, excessive dryness, excessive moisture, or waterlogging, all of which can cause flower drop, fruit drop, and leaf drop. Inadequate ventilation and excessive humidity in greenhouses can also cause chili pepper flowers to fail to pollinate properly and fall off. ④ Pests and diseases. Pepper viral diseases, anthracnose, and mosaic disease (early blight) can easily cause flower and fruit drop. Pepper *Phyllosticta* blight, anthracnose, mosaic disease, leaf spot, and viral diseases can lead to leaf drop. Tobacco budworms and cotton bollworms can also cause fruit drop.

Preventive strategies: ① Select resistant varieties with strong disease resistance and stress tolerance (tolerant to high and low temperatures, etc.). ② Strengthen fertilizer and water management, apply nitrogen, phosphorus, and potassium fertilizers appropriately, and avoid excessive or insufficient nitrogen fertilizer. ③ Plant at an appropriate density, prune in a timely manner, and strengthen ventilation and moisture management when cultivated in a controlled environment to keep good ventilation and light transmission conditions. ④ Apply hormones in early spring during low-temperature seasons. For example, spraying with 40–50 mg/L of anti-falling drug (PCPA) can prevent flower drop and increase early yields. ⑤ Strengthen prevention and control of pests and diseases.

「**Expanded Knowledge**」

Medicinal Functions of Chili Peppers

Fruits: Regulate spleen and stomach functions with warmth, dispel cold, strengthen the stomach, and aid digestion. They are used for stomachaches due to cold, gastrointestinal bloating, and indigestion. They can also be used externally to treat chilblains, rheumatic pain, and lumbar muscle pain.

Roots: Promote blood circulation for detumescence. It can be used externally to treat chilblains.

「**Reviewing and Thinking Questions**」

I. Explanation of Terms

1. Determinate growth of tomato.
2. Indeterminate growth of tomato.
3. Parthenocarpy in cucumber.

II. Gap Filling

1. For tomatoes, they are recommended to harvest during the________ period for ripening, and for fresh consumption, they are recommended to harvest during the ________period.

2. The growth regulator commonly used for preventing excessive seedling growth in tomatoes is ________, and the growth regulator commonly used for flower and fruit retention is ________.

3. Tomatoes can be divided into two types based on plant growth habits: ________ and ________.

4. The growth regulators used for flower and fruit retention in tomatoes include Fanqieling. Its recommended application method is________, and the recommended concentration is________.

5. When flower and fruit retention is required for tomatoes in protected cultivation, the general concentration of 2,4-D solution used is________, and the concentration of ethephon used for ripening of fruits on plants is ________.

6. Tomatoes should be watered evenly, otherwise it may lead to________.

7. During the flowering and fruiting period of tomatoes, poor pollination and fruit setting may occur when the daytime temperature is________and________.

8. ________ pruning is generally used for tomatoes cultivated in a controlled environment.

9. The fruit developed from the flowers at the first branching junctions of eggplant is called________.

III. Choice Questions

1. The cause of blossom end rot in tomatoes is (　　).

A. Calcium deficiency　　B. Magnesium deficiency

C. Boron deficiency　　D. Zinc deficiency

2. The concentration of 2,4-D used to prevent flower and fruit drop in tomatoes is (　　).

A. 10–20 ppm　　B. 20–30 ppm

C. 30–40 ppm　　D. 40–50 ppm

3. Grafting can prevent and control (　　) in eggplants.

A. *Verticillium* wilt　　B. Phytophthora rot

C. Brown spot disease　　D. Yellow tea mite

4. The most suitable period for tomato ripening is (　　).

A. Mature green stage　　B. Turning stage

C. Mature stage　　D. Fully ripe stage

5. Tomatoes cannot naturally turn red at temperatures below (　　).

A. <15°C　　B. <20°C　　C. <25°C　　D. <28°C

6. The optimal concentration of Fanqieling to prevent flower and fruit drop in tomatoes is (　　) ppm.

A. 5–10　　B. 15–20　　C. 30–50　　D. 70–100

7. Tomatoes intended for local sales should be harvested during the (　　) stage.

A. Mature green stage　　B. Turning stage

C. Firm ripe stage　　D. Fully ripe stage

8. To ensure uniformity of fruit size, it is recommended to leave (　　) fruits per cluster for large-fruited tomatoes.

A. 1–2　　B. 3–4　　C. 5–6　　D. 8–10

IV. Thinking and Answering

1. Briefly explain the causes of flower and fruit drop in tomatoes and their prevention methods.

2. How many stages are there in the ripening period of tomatoes, and what are the characteristics of each stage?

3. How can plant regulation be carried out for solanaceous vegetables?

4. How to prevent flower drop, fruit drop, and leaf drop in chili peppers?

Learning Context 3　Production of Legume Vegetables

「**Learning Objectives in This Context**」

Understand the general cultivation characteristics of legume vegetables. Master the biological characteristics, variety types, and cultivation seasons of common beans, cowpeas, and edamame. Acquire the techniques for early cultivation of common beans, cowpeas, and edamame in greenhouses.

「**Analysis of Tasks in This Context**」

Master the tasks involved in seed processing, sowing and seedling raising, land preparation, mulching film covering, transplanting, plant regulation, fertilization and irrigation management, and flower and fruit retention, as well as pest and disease control for common beans, cowpeas, and edamame.

「**Introduction**」

In our daily life, the plant proteins we consume mainly come from legumes, including soybeans, cowpeas, and kidney beans. They are one of important oil crops (vegetables) and are widely planted and researched due to their ability to reduce nitrogen fertilizer usage by hosting rhizobia.

Overview

I. Types of Legume Vegetables

Legume vegetables refer to cultivated populations in the Fabaceae family that are consumed as vegetables, primarily using tender pods or seeds. They include 11 species from 9 genera: *Phaseolus vulgaris* (common bean, lima bean, kidney bean, and runner bean), *Vigna unguiculata* (cowpea), *Glycine* (edamame), *Pisum sativum* (pea), *Vicia* (broad bean), *Canavalia* (vining sword bean), *Lablab purpureus* (hyacinth bean), *Psophocarpus tetragonolobus* (winged bean), and *Mucuna* (velvet bean) (Fig. 3-1). Legume vegetables have high protein content and are rich in nutritional value. They all have papillionaceous corolla which is self-pollinated and easy to preserve seeds, and a deep taproot system that can fix atmospheric nitrogen with root nodules. They require less nitrogen but more phosphorus and potassium in soil nutrition, with good soil drainage and ventilation, and a pH range of 5.5–6.7.

Fig. 3-1 Different Legume Vegetables

II. Biological Commonalities of Legume Vegetables

Except for peas and broad beans, which are long-day plants and prefer cold climates, others are short-day plants that prefer warmth and cannot tolerate cold. As long as the temperature is suitable, they can be planted all year round. Most legume vegetables do not strictly require photoperiods.

Legume vegetables have a long supply period. For example, in the Yangtze River basin, peas, and broad beans are available from March to May; common beans are available from May to June; cowpeas, hyacinth beans, and edamame are available in the hot summer; common beans, cowpeas, sword beans, and hyacinth beans are available from August to November. In southern China, they can be produced all year round.

III. The General Cultivation Principles of Legume Vegetables

Legume vegetable seeds are large and have developed root systems, but are prone to corking and have poor regeneration ability after injury. Therefore, direct sowing or seedling cultivation with root protection measures is recommended. They are relatively drought-tolerant, require good soil drainage and ventilation, and are not tolerant to salt and alkali, with a pH range of 5.5–6.7. They should not be grown continuously in the same plot and should be rotated with non-leguminous crops every 2–3 years. Their root system symbiotically associates with root nodules. When cultivated, they require less nitrogen fertilizer than other crops and more phosphorus and potassium fertilizers.

Sub-context 1 Production of Cowpea

Cowpea, also known as yardlong bean, snake bean, long bean, asparagus bean, Chinese long bean, or pea bean, is a subspecies of the common cowpea plant (*Vigna unguiculata*) in the Fabaceae family. It is native to the tropical regions of Southeast Asia. It is an annual twining herbaceous plant that provides both tender pods and mature seeds as food. They are nutritious, with fresh pods containing protein, lipids, carbohydrates, dietary fiber, calcium, phosphorus, iron, carotene, etc., and can be eaten fresh or processed.

I. Biological Characteristics

1. Botanical features

(1) Roots: Cowpea is a deep-rooted plant with a developed root system, and the taproot can reach a depth of 60–100 cm. They have a weak ability to regenerate from a single root. Their roots are prone to suberization and have poor regeneration ability, with symbiotic nodules, but the rhizobia is not developed.

(2) Stems: Cowpeas can be divided into three categories by growth habits, namely trailing, semi-trailing, and dwarf, and they exhibit a right-handed twining growth habit. For trailing varieties, the terminal buds on both main stem and lateral branches are leaf buds that can grow indeterminately, with a long growth period and requiring trellises for cultivation. For dwarf varieties, the stem is erect, and after the growth of several nodes on the main stem and lateral branches, the terminal bud differentiates into a leaf bud, and the branches grow in a clustered pattern, with a short growth period and no need for trellises. Semi-trailing varieties are intermediate between the two.

(3) Leaves: When germinating, the cotyledons emerge from the soil, and after emergence, the cotyledons fall off as nutrients are depleted. The first two true leaves are arranged in an opposite, simple leaf pattern. The following true leaves are ternately compound leaves that are arranged alternately. The leaflets are smooth, with entire margins and no hairs.

(4) Flowers: The inflorescence is a raceme, with a papillionaceous corolla that can self-pollinate. The flower stalks emerge from the leaf axils of the main stem and have long pedicels. The height of the nodes varies with the variety and growing season, generally ranging from the second to ninth nodes. Lateral branches can also sprout flower branches at the first and second nodes. Each inflorescence has 4–6 pairs of flower buds, often blooming in pairs and podding.

(5) Fruits: The pods are linear in shape, with some pods being curved and strip-

shaped. The length and color vary greatly depending on the variety. They can grow from 20–100 cm long and come in dark green, green, greenish-white, and purplish-red colors. Each pod contains 8–20 or more seeds.

(6)Seeds: The seeds have no endosperm and are shaped like elongated kidneys. The seed coat can be purplish-red, brown, white, black, or mottled. The thousand-grain weight of seeds is 100–150 g.

2. Requirements for environmental conditions

(1) Temperature: Cowpeas prefer warm and heat-tolerant conditions, but are not tolerant of low temperatures and will wither when exposed to frost. Seed germination requires a temperature range of 25–35°C, with the best germination rate and vigor at 35°C. The minimum temperature for germination is 10°C, and the suitable temperature for plant growth is 20–25°C. Growth and development are inhibited above 35°C or below 15°C. The optimal temperature during flowering and podding is 20–30°C, with around 25°C being the most suitable.

(2) Light: Cowpeas can be divided into two categories based on their responses to day length. Some varieties are not sensitive to day length and can grow and develop normally in both long and short daylight seasons. Most yardlong bean varieties belong to this category. The other category has stricter requirements for day length and is suitable for cultivation in short daylight seasons. If cultivated in long daylight seasons, the stems will grow excessively, which will delay the flowering and podding. Day length can affect branching habits and the position of inflorescences. Short day length can promote the emergence of lateral shoots from the base nodes of the main stem, and lower the position of the first inflorescence node. Long day length significantly makes the position of the lateral branch and the first inflorescence on the main stem higher. Cowpeas are light-loving crops and require sufficient sunlight during the growth period. Insufficient light caused by overcrowded planting or excessive growth of stems and leaves can lead to flower and pod drop.

(3) Water moisture: Cowpeas require an adequate amount of water, but can tolerate drought and are not tolerant of waterlogging. A relative air humidity of

70%–80% is suitable, and high temperature and low humidity are the main causes of flower and pod drop. Excessive moisture during the germination and seedling stages should be avoided to prevent reduced germination rate, excessive growth of seedlings, root rot, and seedling death. During the flowering and podding period, appropriate air humidity and soil humidity are required. Excessive rainfall, high humidity, or exposure to dry and cold winds can easily result in flower and pod drop. Excessive soil moisture is detrimental to root and rhizobia activity, and can lead to root rot and disease, resulting in flower and pod drop. High temperatures and dry conditions during the flowering and podding period can also lead to flower and pod drop.

(4) Soil nutrition. Cowpeas have a wide adaptability to soil, and they can generally be cultivated in well-drained and loose soil and grow well in the soil with a pH range of 6.2–7. Since rhizobia are not well-developed, phosphorus and potassium fertilizers should be applied to promote their growth during the seedling stage, while nitrogen fertilizer should be provided appropriately to boost the seedling growth. After flowering and podding, increased amounts of nitrogen, phosphorus, and potassium can be applied to promote flower development and retain the pods.

II. Types and Varieties

There are many varieties of cowpeas. Cultivated cowpeas can be consumed as vegetables and grains. Vegetable cowpeas are classified as early-maturing, medium-maturing, and late-maturing varieties based on the timing of the node position where the first inflorescence grows. Based on their purposes, they can also be roughly divided into hard pod types, which are harvested for mature seeds, and soft pod types, which are used for vegetables. Commonly cultivated cowpeas include yardlong beans and catjang. Based on growth habits, they can be classified into trailing, semi-trailing, and dwarf, with semi-trailing varieties being rarely cultivated in China. In China, cowpeas can be classified into three categories based on pod colors: green pod, white pod, and red (or purple) pod.

(1) Green pod (also known as green string bean): They have thin stems, small

leaves, dark green color, as well as long and slender pods in green. The pod flesh is tender, thick, and crispy, and of good quality. They can tolerate slightly lower temperatures, but have slightly poor heat resistance, shorter harvesting periods, and lower yield. They are mainly suitable for cultivation in spring and autumn. The main varieties include Tiexianqing, Xiyeqing, and Zhuyeqing in Guangdong, Qingdoujiao and Zaoqinghong in Zhejiang, Qingfeng and Datiaoqing in Shandong, Chaoyangxian in Guizhou, etc.

(2) White pod (also known as asparagus beans): They have thicker and larger stems, and larger and thinner leaves, which are green in color. Their pods are relatively large, light green or greenish-white in color, with thin and loose flesh, and the seeds are easily visible. They have strong heat resistance and higher yield. They are mostly suitable for planting in spring and summer. Main varieties include Zhijiang 28–2, Qiujiang 512, Yangjiang 40, Gaochan No.4, as well as Hongzuiyan which are distributed in various regions.

(3) Red pod (also known as purple yardlong bean): They have thicker and stronger stems, purplish-red color between stems and petioles, and larger green leaves. Their pods are purplish-red, relatively thick and short, with medium tender flesh, and are prone to aging, so they have a shorter harvesting period, and lower yield. The main varieties include purple yardlong beans in Shanghai and Nanjing, Xiyuanhong in Guangdong, Hongshanyugugu, Zijia, and Bailu in Hubei, Zijiang in Beijing, etc.

III. Cultivation Season and Methods

Due to differences in heat resistance and responses to day length, the cultivation season and selection of varieties should be determined based on local climate conditions. In regions south of the Yangtze River, cowpeas can be cultivated in spring, summer, and autumn, with a long production season. In spring, cowpeas are mostly sown in open fields from March to May, while in autumn, sowing takes place from July to August. For early spring cultivation, the cowpeas are sown or their seedlings are cultivated in late February to late March, transplanted from late March to late April, and harvested from May to August. In South China, seedling raising can start as early as February, and the seeds can be sown until early September. In northern regions, cowpeas are mainly sown from early April to mid-May.

IV. Cultivation Techniques

1. Strong seedling cultivation

Cowpeas are generally directly sown. To ensure early planting and prevent seed rotting or seedling death caused by low temperatures and rainfall, especially in early spring, insulated seedbeds or nutrient soil blocks (nutrient bowls) can be employed for seedling raising. Sowing and seedling raising usually take place in late March in Jiangsu and Zhejiang, and in February in South China. Before seedling raising, select high-quality seeds and water the soil thoroughly. Sow 2–3 seeds per bowl, covered by a greenhouse to maintain warmth and moisture. After emergence and before transplanting, it is better to keep the temperature around 20°C.

Seedlings should be stored for 4–5 days before transplanting. Transplanting is generally carried out before the first compound leaf appears. Transplanting of seedlings that are raised in the nutrient bowls can be appropriately delayed. Seedlings should be transplanted on sunny days. Cowpeas are sown directly, with spring sowing from March to May and autumn sowing from July to August in South China. In the Yangtze River basin, autumn sowing takes place from late June to early August. The plant and row spacing for direct sowing is (50–60) cm × (20–40) cm, with 2–3 plants per hole. Approximately 6,000 holes are needed per mu, requiring about 1.2 kg of seeds.

2. Field management

(1) Fertilizer and water management. For base fertilization, 2,000–3,000 kg of well-rotted farmyard manure and 40 kg of microbial compound fertilizer are applied per mu respectively in ridge tillage and soil cultivation. Topdressing should be applied lightly and frequently. Generally, topdressing is applied once every 7–10 days. Before the setup of the bamboo trellis, a 1% urea solution is suitable, and after the setup of the bamboo trellis, the concentration can be increased. Heavy fertilizer,

approximately 20 kg of compound fertilizer per mu, should be applied after podding. Water moisture should be controlled during the flowering period, and the soil should be kept moist after podding. Watering is also necessary at noon during intense sunlight to cool down the plants.

(2) Trellising and vine direction. When seedlings reach a height of 20–25 cm, trellising should be done with a herringbone trellis to prevent attack by typhoons and heavy rain and reduce lodging. Vine direction should be done on sunny afternoons when the branches are less likely to break.

(3) Plant regulation (pruning). ① Remove all lateral buds below the first inflorescence of the main vine. ② Pinch off lateral branches that grow on the upper part of the main vine during the mid- to late growth stage. ③ Top off the main vine when it reaches about 2 m to promote the formation of secondary flower buds of each inflorescence on the lateral branches.

3. Pest and disease control

Yardlong beans are susceptible to diseases such as leaf mold, root rot, blight, rust disease, legume pod borer, cutworm, thrips, leafminers, and mites, which should be prevented and controlled in a timely manner.

4. Harvesting

During harvest, we should avoid damaging other flower buds on the inflorescences and do not remove the pedicel along with the inflorescence. We should hold the base of the pod and gently twist it left and right before picking. For cowpeas planted in spring, it is recommended to harvest the marketable pods 11–13 days after inflorescence, while for cowpeas planted in summer, they should be harvested 9–11 days after flowering.

V. Common Cultivation Issues and Their Prevention Strategies

1. Summer slump

After the first peak yield in production, cowpea plants tend to age early, with a decrease in the number of flowering and podding, which results in a significant drop in yield. This phenomenon generally occurs during the hot summer, so it is often referred to as "summer slump". The main reasons for the occurrence of the "summer slump" are: After a large amount of nutrients is consumed during the first peak of yield, the fertilizers and water are not replenished in time, which leads to premature defoliation. Inadequate pruning and pinching result in poor ventilation and light transmission. High temperatures, drought, or excessive rainfall cause severe leaf shedding. Pests and diseases damage the functional leaves. Prevention and control measures: Application of sufficient base fertilization to prevent nutrient deficiency caused by late topdressing, and multiple topdressing during the flowering and podding stage. Timely pruning and pinching. Timely water supplementation during hot seasons and proper drainage after heavy rain. Strengthening pest and disease control.

2. Early leaf shedding

During the peak harvest period of cowpeas, sometimes a large amount of leaf shedding weakens photosynthesis and leads to reduced growth vigor and ultimately decreased yield. Sometimes, cowpeas may also experience leaf shedding after emergence or transplanting in spring. The reasons for leaf shedding during the seedling stage of spring cowpeas: Low temperatures in early spring may lead to poor root system development, which inhibits growth. Low-quality transplanting and prolonged seedling acclimation result in yellowing and shedding of lower leaves. After emergence, adverse conditions such as low temperatures and drought prevent sufficient nutrient supply to the cotyledons, and lead to leaf shedding. The reasons for leaf shedding during the peak harvest period: Excessive rainfall or drought causing waterlogging, and nutrient deficiency leading to premature nutrient depletion, as well as damage from diseases and pests. Preventive measures: Avoiding early sowing, strengthening the water management during the podding stage, proper fertilization after harvest begins, and timely prevention and control of pests and diseases.

「Summary of Sub-context」

This sub-context mainly discusses the biological characteristics and cultivation techniques for cowpeas. In terms of cultivation techniques, it is important to focus on the field management of cowpeas.

「Expanded Knowledge」

Nutritional Analysis of Cowpeas

(1) Cowpeas provide digestible and absorbable quality protein, appropriate amounts of carbohydrates, multivitamins, micronutrients, etc., which can supplement the signature nutrients of the human body.

(2) Vitamins B contained in cowpeas can maintain normal digestive gland secretion and gastrointestinal motility functions, inhibit choline activity, help digestion, and increase appetite.

(3) Vitamin C contained in cowpeas can promote the synthesis of antibodies and improve the body's antiviral effect.

(4) Phospholipids of cowpeas can promote insulin secretion and participate in sugar metabolism. It is an ideal food for people with diabetes mellitus.

Sub-context 2 Production of Kidney Bean

Kidney bean, also known as *Phaseolus vulgaris*, jade beans, etc., is native to Central and South America. Belonging to the Fabaceae family, Kidney bean is an annual herbaceous plant, whose tender pods and dried beans are edible. Its nutritional value is high. Each 100 g of fresh kidney bean contains 1.5 g of protein, 0.2 g of fat, 4.7 g of carbohydrates, 0.8 g of crude fiber, 44 mg of calcium, 39 mg of phosphorus, 1.1 mg of iron, 0.24 mg of carotene, and 0.08 mg of vitamin B1, and 9 mg of vitamin C; Its dry seeds contain 23.5% protein and 50.6% carbohydrate.

I. Biological Characteristics

1. Morphological characteristics in botany

(1) Roots: The roots of the kidney bean are prone to subrization, with a weak regeneration ability, non-obvious taproot, and lateral root system developing faster

than the above-ground part, and developed rhizobia.

(2) Stems: Its stems are herbaceous, multiangular, with hairs on the surface, slender and green. Its stems grow in a left-handed twining manner, with strong branching ability. According to its growth habits, they can be divided into three varieties: dwarf variety (with determinate growth), vining variety (with indeterminate growth), and semi-vining variety. Dwarf variety has upright stems and short internodes, and is suitable for early-maturing cultivation. Vining variety has a main vine of 3–4 m, with short primary internodes. Later, the internodes of the stem nodes are elongated. Side branches and inflorescences can grow at each axillary bud of stem nodes. Stems grow in a left-handed twist upward, and trellises are needed for the cultivation.

(3) Leaves: When the cotyledons of the kidney bean grow above the surface, the first pair of true leaves are opposite single leaves in a heart shape, while the second and subsequent leaves are alternate, ternate compound leaves.

(4) Flowers: The kidney bean has raceme inflorescences, each with 5–6 flowers, up to more than 10 flowers. It has papillionaceous flowers, with corollas in white, yellow, purple, red, pink, etc. The plant is self-pollinated, but its outcrossing rate is 0.2%–10%.

(5) Fruits: Kidney bean pods are mostly in pairs and in round or oblate shape. Their outer skins have green, yellow, red, or slightly purple patterns.

(6) Seeds: Kidney bean seeds are large and kidney-shaped, and their seed coats have white, black, yellow, purplish-red, and brown patterns. Their thousand-seed weight is 300–600 g. Its seeds have a time limit of germination of 2–3 years.

2. Requirements for environmental conditions

(1) Temperature: Kidney beans prefer warm temperatures but cannot tolerate frost, and dwarf varieties have stronger low-temperature tolerance than vining varieties. Kidney beans have an optimum temperature for germination of 20–25°C, and are hard to germinate above 35°C and lower than 8°C. The optimum temperature for its seedling development is 18–20°C. If the temperature is lower than 13°C, root nodules will not form on the roots. The optimum temperature for kidney beans' flower bud differentiation is 20–25°C, and incomplete flowers will

easily appear if it is lower than 15°C or higher than 27°C. The optimum temperature for their flowering and podding is 18–25°C. Too high or too low temperatures, will affect the number of pods and the number of seeds in the pods.

(2) Light: Kidney beans are less sensitive to sunlight than other legumes. Most of them do not have strict requirements on the length of sunshine, so they can be induced and cultivated in spring, summer, and autumn in various places in both North and South China.

(3) Water moisture: Kidney beans are highly resistant to drought, but excessive drought or waterlogging is not conducive to root system growth. The suitable soil humidity during the growth period is 60%–70% of the maximum water capacity of the field, and the suitable relative air humidity during the flowering and podding stages is 80%–90%.

(4) Soil nutrition: Kidney beans are suitable for cultivation in neutral soil with rich humus, deep soil layer, and good drainage, with an optimum pH range of 6.2–6.8. The absorption of nitrogen, phosphorus, and potassium by the bean plant increases significantly as the pods form. After the flower bud differentiation, nitrogen promotes plant growth, increases the number of flowers, and produces more pods; however, too much nitrogen can lead to flower drop.

II. Types and Varieties

Kidney beans can be divided into two types based on the degrees of fibrosis of pods: soft pod variety (used as vegetable) and hard pod variety (used as foodstuff). According to the color of the pods, they can be divided into green, yellow, red, and purple varieties. According to the color of the seeds, they can be divided into black, white, red variety, yellowish-brown, and piebald varieties. In terms of production, they are divided into dwarf (ground beans) and vining varieties (trellis beans) based on their growth habits. A few intermediate types are called semi-vining varieties.

1.Vining variety

It is also known as trellis beans, with main vines of 2–3 m or longer, long internodes and left-handed climbing growth habit. It belongs to the indeterminate production plant. Its axillary buds at all stem nodes can have lateral branches or inflorescences, and generate flowers and pods successively. Its flowering stage is long and the pods mature late.

2.Dwarf variety

It is also known as ground beans or squatting beans. Its plant is dwarf and upright, with a height of 30–50 cm. Its basal internodes are short and the upper internodes are slightly longer. When the main stem has 4–8 nodes, the terminal buds become floral buds. Dwarf variety of kidney beans belongs to the determinate growth variety. Its growth period is short, with early flowering and maturity, a period of 40–50 days from sowing to first harvest, and the harvesting period is 15–20 days. Its yield is low and the quality is poor.

At present, there are many excellent varieties cultivated in China. For example, those with larger planting areas in the North and South China include Yixuan No. 1, Jiadouwang (Trellis bean king), Yundouwang, Shuangfeng No. 2, Shuangfeng No. 3, Chulyu No. 2, Taiwan No. 2, Wujin No. 2, Xinyin No. 2, Kaluote, Kapeng, Winner (77–10), Yunfeng (623), Zhejiang Aizao No.18, Nongyou Zaosheng, Kanghan No. 2, Qingdaodou No. 27, Bilong No. 1, Changcaidou No. 1, Gusu Zaodi Sijidou, etc.

III. Cultivation Season and Methods

Direct sowing in spring can be performed in early to mid-March. Low-temperature, rainy weather often occurs in the Yangtze River basin, which leads to rotten bean seeds and dead seedlings in spring. Therefore, seedling culture and transplantation are often adopted. Generally, seedlings are grown from late February to early March for early spring cultivation. This can extend the growing season, bring the beans to market earlier, and increase yields. Direct sowing in spring in South China is performed from January to February. Autumn cultivation is generally performed by direct sowing. The sowing time in the Yangtze River basin is from mid-July to early August, and in South China it is from September to October.

IV. Cultivation Techniques

1. Strong seedling cultivation

Large, plump seeds free of diseases and insects should be selected before sowing, and insoluble for 1–2 days, which is conducive to unified germination. In order to prevent diseases, seeds can be soaked in 1% formalin for 20 minutes, then rinsed with clean water before sowing; Thiram of 0.3% of the seed weight can also be added before sowing. For direct sowing, 3–4 beans should be sowed per hole. The row spacing of the dwarf variety is 33–40 cm, and the hole spacing is 20–26 cm; the row spacing of the vining variety is 65–85 cm, and the hole spacing is 20–27 cm; two rows of seedlings are planted in each saddle and share one trellis. The amount of seed used for each mu of land is 2–3 kg.

Seedlings can be cultured in row covers or cold frames, and nutrient pots or nutrient soil blocks can be used for hole sowing in the frame. In spring, 5–6 kg of seeds can be sown in 5 m^2 for culturing seedling in cold frames. After the seedlings emerge, remove the film, ensure proper ventilation, and transplant the seedlings with soil when they are 15–20 days old.

2. Field management

(1) Fertilizer and water management. Before sowing, sufficient base fertilizer should be applied, 1,000–1,500 kg of decomposed farmyard manure, 30–40 kg of compound fertilizer, and 20–30 kg of lime should be applied to 1 mu; Topdressing should be performed following the principle of "light dressing at first and heavy dressing later", that is, topdressing should be started after the true leaves appear, 2.5 kg of urea should be applied 1–2 times per mu, and topdressing of 3–4 times is needed during the flowering and podding stage, with 10–15 kg of compound fertilizer and 10 kg of potassium fertilizer be applied per mu each time. In addition, 0.3% solution of potassium dihydrogen phosphate should be sprayed on stems and leaves every 6–7 days during the flowering stage, application of 5–25 mg/kg gibberellin to the top of stems can also effectively prevent the falling of flowers and pods and increase the yield. The drainage and irrigation of kidney beans are the same as that of cowpeas.

They should be watered frequently to keep wet, and drainage is needed after raining.

(2) Supplementary seedling, intertilling, trellising, and directing vines. Seedlings that are missing or have injured or diseased basal leaves should be replaced in time. Loosening the soil by intertillage can quickly increase ground temperature and soil permeability, promoting root growth and rhizobia development. The first intertillage weeding can be done after direct sowing and full germination, and the transplantation of cultured seedlings can be performed after seedling adaption. The second intertillage weeding can be done before budding, or before the sprouting of vines of vining beans. Vining beans are generally trellised before sprouting of vines, and herringbone trellis are preferred. After trellising, vines should be directed manually once.

(3) Pest and disease control. The diseases and insect pests of kidney beans are similar to those of cowpeas. For prevention and control methods, please refer to the section "Production of Cowpea". In addition, kidney beans are susceptible to anthracnose. For their prevention and treatment, 500-fold solution of 75% wettable chlorothalonil powder, 300-fold solution of 50% anthrax thiamine wettable powder, etc. can be sprayed.

3. Harvesting

Under normal circumstances, the tender pods can be harvested 10–15 days after flowering. Under low temperatures, they can be harvested about 15–20 days after flowering. They can be harvested when the pods change from flat to round, the color changes from green to light green, the surface is shiny, and the seeds slightly bulge or do not yet bulge.

V. Common Cultivation Issues and Their Prevention Strategies

1. Flower and pod dropping

There are many reasons for the yield reduction of kidney beans, among which falling flowers and pods are the fundamental reasons. Kidney beans have a high degree of flower bud differentiation and many flowers, but with a very low podding rate.

The main reasons for flower and fruit drop are as follows. The first one is nutritional factors. In the early flowering stage, due to the simultaneous vegetative growth and reproductive growth, the inflorescence cannot receive sufficient nutrition. In the middle stage, nutritional imbalance occurs due to competition between inflorescences, between flowers within one inflorescence, and between flowers and pods. In the later stages of growth, the plant's weakness and poor environmental conditions can also cause flower drop. The second one is environmental conditions. Temperatures higher than 30°C or lower than 15°C during the flowering and podding stage will affect pollination and cause flower drop. During the flowering stage, rain or high temperatures and drought can influence pollination and lead to flower dropping; insufficient light and low photosynthetic products result in poor development of flower organs and falling of flowers. The third one is cultivation management. Too early watering during the initial flowering stage will cause the plant to enter the vegetative and reproductive growth stages prematurely, and the conflict between stem and leaf growth, and flowering and podding will become more prominent. Too much nitrogen fertilizer applied in the early stage will result in excessive vegetative growth and limited flower bud differentiation; insufficient fertilizers may contribute to malnutrition; too high planting density and untimely pruning can result in poor ventilation and poor light transmission; overdue harvesting can result in over-consumption of nutrients by pods; serious pests and diseases, etc. will cause flower and pod drop.

Measures to prevent flower and pod drop: The first is to choose excellent varieties with high podding rates. The second is to sow in due time to reduce or avoid hazards from high and low temperatures. Protection facilities or reasonable intercropping and relay cropping can also be used to improve microclimate conditions. The third is to strengthen field management and rationally regulate the balance between vegetative growth and reproductive growth. For example, reasonable dense planting, timely trellising and vine direction, appropriate application of nitrogen fertilizers, increased application of phosphorus and potassium fertilizers, controlled watering during flowering, timely pruning and topping, prevention and control of diseases and insect pests, and timely harvesting. The fourth is to spray 5–25 mg/L naphthylacetic acid or 2 mg/L p-chlorophenoxyacetic acid (commonly known as anti-dropping agent) during the flowering stage, which can reduce flower and pod drop.

2. Premature aging of fruit pods

Kidney beans take tender pods as their edible organs, and pod aging greatly reduces their quality. Pod aging is mainly related to varieties and environmental factors. Fiber-free varieties are not prone to aging. Among environmental factors, high temperatures exceeding 31°C or daily average temperatures exceeding 25°C are most likely to cause pod aging. In addition, malnutrition and lack of water can also promote fiber formation.

The first measure to prevent premature aging of pods is to choose anti-aging varieties; The second measure is to sow seeds in due time to avoid podding in high-temperature seasons while strengthening water and fertilizer management; The third measure is to harvest pods in time before they age.

「Summary of Sub-context」

The focus of this sub-context is to master the cultivation techniques of kidney beans, as well as the causes of flower and fruit drop and their prevention and control methods.

「Expanded Knowledge」

Nutritional and Medicinal Values of Kidney Beans

(1) Kidney beans are rare food with high-potassium, high-magnesium, and low-sodium, which endow them with great use in nutritional therapy. Kidney beans are especially conducive for patients with heart disease, arteriosclerosis, hyperlipidemia, hypokalemia and those who avoid salt.

(2) Modern medical analysis believes that kidney beans also contain unique ingredients such as saponins, uremic enzymes, and various globulin, which can improve the body's immunity, enhance disease resistance, and activate lymphatic T cells, promote the synthesis of DNA and other functions, and have an inhibitory effect on the development of tumor cells, so they have attracted the attention of the medical community. The urease contained in them is very effective for patients with hepatic coma.

Sub-context 3 Production of Vegetable Soybean

Vegetable soybeans, also known as edamame, are native to China. Vegetable soybeans belong to *the Glycine* genus of the Fabaceae family. They are annual herbaceous plants, and their edible parts are mainly immature fresh seeds. It is rich in nutrients. Every 100 g of tender seeds of edamame contains 13.6–17.6 g of protein, 5.7–7.1 g of fat, 31–43 g of dry matter, 100 mg of calcium, 219 mg of phosphorus, 6.4 mg of iron, and 0.2 mg of carotene.

I. Biological Characteristics

1. Botanical features

(1) Roots: Vegetable soybean has a well-developed root system and a strong taproot. Its lateral roots are well-developed, which can extend by 40–50 cm in all directions and reach a depth of 1 m. Its rhizobia are also well developed and mainly distributed in the cultivated layer within a depth of 20 cm.

(2) Stems: Its stems are upright and strong, with irregular edges and corners, and have hairs on them. Its tender stems are green and purple. Generally, green stems have white flowers and purple stems have purple flowers. Old stems are grayish-yellow or brown.

(3) Leaves: When the cotyledons appear, there is one pair of primary leaves,

which are alternate, followed by alternate, ternately compound leaves. Its leaflets are ovate or elliptical, and the leaf surface is covered with hairs or hairless.

(4) Flowers: Vegetable soybeans have short racemes, which grow between the leaf axils of nodes or at the terminals. Each inflorescence can bear 3–5 pods and is self-pollination. Vegetable soybeans have small, white, lavender or purple flowers.

(5) Fruits: Vegetable soybeans have rectangular and flat pods, which are densely covered with white or brown hairs. Each pod contains 1–4 seeds.

(6) Seeds: Vegetable soybeans have different sizes, shapes, and colors from variety to variety. Their seeds are elliptical, spherical, and oblate, and in yellow, green, dark brown, and spotted bi-color. Their seed umbilicuses are yellowish-white, purple, black, or brown, indicating their varieties. Their thousand-seed weight is 100–500 g.

2. Requirements for environmental conditions

(1) Temperature: Vegetable soybeans like warmth and cannot tolerate frost. Their seeds begin to germinate at 10–11°C, with an optimum temperature at 20–22°C. Their seedlings can tolerate short-term low temperatures. Their optimum temperature during the growth period is 20–25°C. Their flower bud differentiation stage is 25–30 days, and the daytime temperature of 24–30°C and nighttime temperature of 18–24°C are conducive to flower bud differentiation. The optimum temperature for flowering and podding is 22–25°C. The optimum temperature for the period from young pods to mature seeds is 19–20°C, during which they are particularly sensitive to temperature, and high temperatures will stop growth prematurely.

(2) Light: Vegetable soybeans are short-day crops, and their responses to day length vary depending on the varieties. Generally speaking, determinate growth types and early-maturing varieties in South China do not have strict requirements on lighting length and can be cultivated in both spring and autumn. The determinate growth types and late-maturing varieties in North China are short-day crops. Therefore, northern varieties that moved to South China tend to bloom earlier, while southern species that moved to North China tend to flower later.

(3) Water moisture: Vegetable soybeans are leguminous vegetables that require much water, but have poor waterlogging tolerance. Water requirements vary with

growing periods. When their seeds germinate, enough water allows quick and full emergence. The water content of the field should be maintained at 60%–65% of the maximum water-holding capacity in the seedling stage, 65%–70% in the branching stage, 70%–80% in the flowering and podding stage, and 70%–75% in the pod filling stage.

(4) Soil nutrition: Vegetable soybeans do not have strict soil quality requirements. However, sandy loam with deep soil, good drainage, and rich organic matter are more conducive to the growth and development of vegetable soybeans, and the pH value suitable for growth is 6.5–7. As for nutrient requirements, fertilizer absorbed in the early stage (from sowing to the flowering stage) accounts for less than 15% of the total fertilizer, and the fertilizer absorbed during the flowering and podding stage accounts for more than 80% of the total fertilizer. To produce 100 kg of grains, approximately 8.5 kg of nitrogen, 0.84 kg of phosphorus, 3.04 kg of potassium, and various micronutrients are required.

II. Types and Varieties

According to their growth habits, vegetable soybeans can be divided into indeterminate growth types and determinate growth types. The indeterminate-growth vegetable soybeans have vine-like stems, small and numerous leaves, and leaf buds at the terminals. They have a long flowering stage and high yields, and are commonly cultivated in Northeast and North China. The determinate-growth vegetable soybeans have upright stems, large and few leaves, and floral buds at the terminals. They have a concentrated flowering stage, mature earlier, and are widely distributed in the Yangtze River basin.

According to the length of the growth period of vegetable soybeans, they are divided into three types: early-maturing, medium-maturing, and late-maturing beams. Early-maturing vegetable soybeans have a growth period of less than 90 days, which are sown in spring in the Yangtze River basin and harvested from late May to late June. Medium-maturing vegetable soybeans have a growth period of 90–110 days and are harvested from early July to early August. Late-maturing vegetable soybeans have a growth period of more than 120 days and are harvested from late September to late October.

China is the origin of vegetable soybeans and has rich germplasm resources. For example, the Heihe series, Tieling series, Fushun series, etc. launched in Northeast China.

Spiced edamame (Zhejiang local variety): Its plants are upright, about 1 m high, late-maturing, with pods about 6 cm long and 1.4 cm wide, green in color. There are 2–3 beans in each pod. The beans are large, have a glutinous aroma, and are of high quality. Its mature seeds are brown, with a yield of 400–500 kg per mu.

Wuyueba (Zhejiang local variety): With a plant height of 55–60 cm, it is an early-maturing variety. Its pods are generally 5 cm long and 1.8 cm wide. Each pod contains about 3 beans. Its fresh beans are yellowish-green, crisp, tasty, and high quality. Its mature seeds are yellowish-white.

Ciguqing (Shanghai local variety): With a plant height of 90 cm, it is a late-maturing variety. Its pods are about 6.2 cm long and 1.4 cm wide. Its pods have 2–3 beans, with a green seed coat and dark brown navel. Its pods and beans are crisp and of high quality. The yield of fresh pods per mu is 700–800 kg.

Baishuidou (Chengdu local variety): With a plant height of 45–50 cm, it is an early-maturing variety. Its pods are about 6 cm long, 1.2 cm wide, and yellowish-green. Its pods have 2–3 beans on average. The beans are large, tender, delicious, and of high quality. Its yield per mu is approximately 750 kg (with poles).

III. Cultivation Season and Methods

In most places south of the Yangtze River, vegetable soybeans can be produced in spring, summer, and autumn. Generally, spring sowing varieties can be sowed or cultured or directly sowed from February to April, summer sowing varieties from April to June, autumn sowing varieties from late June to early August, and autumn sowing varieties in South China from July to August.

IV. Cultivation Techniques

1. Strong seedling cultivation

Vegetable soybeans are generally sown directly. The field plots should be selected and applied with sufficient base fertilizer and an appropriate amount of phosphorus and potassium fertilizer. Poisonous bait should be spread before sowing to prevent underground pests. 1–1.5 kg of 3% Miler granules can be mixed with seeds for every mu. Inoculation with rhizobia before sowing is an effective yield-increasing measure. The field plots can be spread with topsoil containing rhizobia, inoculated with soil leachate, or inoculated with artificially cultivated fine strains. Seeds can be sowed in holes or sowed scattered in furrows, with a nutritional area of (25–35) cm × (16–25) cm, 4–5 seeds per hole, and 2–3 seedlings should be left in each hole. Generally, it is appropriate to grow 25,000–30,000 early-maturing plants, 18,000–20,000 medium-maturing plants, and 15,000–17,000 late-maturing-plants per mu. The weight of seeds used per mu is about 3–4 kg.

Spring sowing in South China often encounters low temperatures and rainy weather, so seedling culture and transplanting can be adopted for early market release. Seedlings can be cultured with row covers or cool frames from late February to early March. Nutrient pots or nutrient soil blocks (7–8 cm) can also be used for hole sowing. Seedlings should be transplanted with soil before the true leaves unfold, that is, when they are 15–20 days old.

2. Field management

(1) Fertilizer and water management. Generally, 1,000–1,500 kg of decomposed compost, 30 kg of calcium superphosphate, or 10–15 kg of compound fertilizer containing phosphorus and potassium may be applied per mu as base fertilizers. The base fertilizers may be mixed and spread on the fields. Thin human feces can be applied as topdressing 2–3 times after the quantities of seedlings are determined to facilitate root taking and germination. If there is potassium deficiency in the late growth period, apply 250 kg of plant ash per mu in the early morning when the dew is still wet, or spray aqueous solution of 0.3% potassium dihydrogen phosphate once every 7 days 2–3 times in total. All these measures have a significant yield-increasing effect.

(2)Topping in due time. The flowers that bloom in the later growth stages of vegetable soybeans often fail to mature in time. Timely topping can inhibit growth, promote maturing, reduce flower drop and pod deflation, and increase yield. Generally, for determinate growth types, if the growth is too vigorous, they can be topped during the early flowering stage. Late-maturing varieties of indeterminate growth types should be topped after the full flowering stage.

3. Pest and disease control

The main diseases of vegetable soybeans include brown spot, black spot, sporulating nematode, sclerotinia rot, root rot, downy mildew, mosaic virus, etc. The main pests include soybean borers, pod borers, Spodoptera litura, red spider mites, etc. All these diseases and pests should be prevented and treated in time.

4. Harvesting

Generally, they should be harvested when the pod shell changes from green to yellow-green, the beans are plump but still green, and there are still seed coats around them. At this time, the sugar content is high and the quality is good. When harvesting, customs vary from place to place. In Sichuan, the leaves are picked and sold together with the pods; in regions south of the Yangtze River, green pods are picked, and sometimes beans are peeled off and harvested 1–3 times. The yield rate of beans is generally 40%–50%.

「**Summary of Sub-context**」

This sub-context focuses on mastering the cultivation technology and the field management technology of the vegetable soybean.

「**Expanded Knowledge**」

Methods of Making Soybean Sprouts

(1) Seed selection: Choose new bean seeds with a high germination rate and good germination potential. Remove broken, insect-eaten, and moldy seeds. Wash with water to remove impurities and floating seeds.

(2) Seed soaking and germination: Soak the seeds in clean water for 8–12 hours, take them out, and drain off the excess water. Put them into a wooden barrel or bamboo basket. There should be drainage holes at the bottom of the barrel. The wooden barrel or bamboo basket should be sterilized in advance, and the beans should be filled to a thickness of 13–16 cm, and then placed in a warm room. During the germination stage, keep the temperature at around 25°C and spray water every 4–5 hours to maintain the humidity of the seeds. Shade the seeds with a clean linen cloth to keep them warm and moisturized. Soybean sprouts can be sold in 5–7 days when the buds are 7–10 cm long.

「**Reviewing and Thinking Questions**」

I. Explanation of Terms

Rest on hot summer days.

II. Gap Filling

1. Kidney beans are generally divided into________ type and ________ type based on their growth habits.

2. The root system of vegetable soybeans has symbiotic ________ and can fix ________ in the air to synthesize nitrogen substances.

3. The flowering and podding stage is the peak time for vegetable soybean plants to absorb ________ and ________.

4. The flowers of cowpeas are ________ flowers and are ________-pollinating crops.

5. Cowpea has few root nodules, appropriate topdressing of________ fertilizers and ________ fertilizers can be applied during the seedling stage to promote the growth of root nodules.

III. Choice Questions

1. Pea requires a large amount of (　　), and extraradical topdressing during flowering and podding can effectively increase yield and quality.

A. Boron　　　　B. Molybdenum

C. Zinc　　　　D. Silicon

2. The legume vegetable that can tolerate freezing climate is (　　).

A. Kidney bean　　　　B. Cowpea

C. Broad bean　　　　D. Vegetable bean

IV. Thinking and Answering

1. Why are legume vegetables often sown with dry seeds? What should you pay attention to when culturing seedlings?

2. What measures should be taken to ensure full germination when directly sowing legumes?

3. What are the causes for the flower and pod dropping of kidney beans? How to prevent and treat it? What are the characteristics of fertilizer and water management for kidney beans?

4. Briefly describe the significance and method of cowpea pruning.

5. What are the causes of cowpeas' early defoliation and later "rest in hot summer days", and how can they be prevented and controlled?

Learning Context 4 Production of Cabbage Vegetables

「Learning Objectives in This Context」

Master the main varieties and common characteristics of cabbage vegetables, the biological characteristics of the main varieties, and the basic knowledge about the production of cabbage vegetables. Be able to correctly select cabbage cultivars according to the cultivation facilities and cultivation seasons. Be able to formulate cabbage production plans and correctly produce cabbage vegetables.

「Analysis of Tasks in This Context」

Master seedling cultivation, fertilizer and water management, and pest and disease control of cabbage vegetables.

「Introduction」

Cabbage vegetables such as Chinese cabbage and cabbage are popular vegetables. They have rich nutrients, high yields, and high economic value, and are deeply favored by vegetable growers.

Overview

I. Classification of Cabbage

Cabbage vegetables refer to plants of the genus *Brassica* of the Brassicaceae family, including three species: brassicas, cabbage, and mustard. *Brassica* species include the turnip subspecies, *Brassica rapa* subspecies, and Chinese cabbage subspecies. The turnip subspecies has obvious petioles, deeply lobed or fully lobed leaves, and enlarged fleshy roots. It is a root vegetable. The cabbage subspecies has open leaves and short plant size. Its most varieties have smooth leaves, with obvious petioles and no obvious leaf wings. The cabbage subspecies includes var. *communis*, var. *rasularis*, var. *tsai-tai*, var. *purple tasi-tai*, var. *tai-tsai*, and var. *multiceps*. The Chinese cabbage subspecies is large in size and has obvious petioles and leaf wings. Among them are loose-leaf variant (var. *dissoluta*), semi-headed variant (var. *infarcta*), flower-center variant (var. *laxa*), and headed varieties (var. *cephalad*). The numbers of chromosomes of the above subspecies and variants are all 2n=2x=20. They have the same basic chromosome set, so the natural hybridization rate between subspecies and variants is high. Cabbage species include head cabbage, cauliflower, Brussels sprouts, collard greens, kohlrabi, kale, and other variants. Mustard includes leaf mustard, stem mustard, root mustard, seed mustard, and other variants (Fig. 4-1).

II. Characteristics of Cabbage

Cabbage vegetables are cold-loving crops with strong cold resistance but weak heat resistance. Its suitable average monthly temperature for cultivation is 15–20°C. Wutaitsai and tai-tsai are the most cold-tolerant among cabbages, while some varieties of common cabbage are more heat-tolerant. After the germination of seeds or under a low temperature of seedlings below 15°C, the vernalization can be completed after a certain period of time. Long daylight and relatively higher temperature (18–22°C) conditions are conducive to bolting, flowering, and seed maturation. They are suitable for autumn cultivation. They have the same or similar

Chinese cabbage Wuta-tsai Mustard

Bok choy Cabbage Cauliflower

Purple cabbage Tsai-tai Broccoli

Fig. 4-1 Different Cabbage Vegetables

biological characteristics, and share many similarities in cultivation techniques.

(1) Cabbage vegetables like mild climates and are suitable for cultivation in seasons with average monthly temperatures of 10–22°C. The monthly average temperature of 25°C is not suitable for their growth. Most of them have strong cold resistance. Among them, Chinese cabbage, stem mustard, cauliflower, etc. are semi-cold-tolerant vegetables and can withstand light frost.

(2) Cabbage vegetables need to complete the vernalization under low temperature and the light stage under long daylight conditions to complete the entire growth cycle. During their cultivation, attention should be paid to their growth and development rules to avoid early bolting, which will affect yield.

(3) They have shallow root systems with weak water absorption capacity.

Their leaves are large, with vigorous transpiration, thus requiring high soil moisture and great air humidity. Timely irrigation is also needed during cultivation.

(4) They grow fast, with a high yield, thus requiring much mineral nutrients and fertile soil. The major fertilizer they need is nitrogen, while they also need the combined use of phosphorus and potassium fertilizers. Nitrogen fertilizer can promote the growth of their leaf clusters and has the greatest impact on their yield and quality. Phosphorus and potassium are beneficial to the growth of the head and the differentiation and development of flower stalks.

(5) This type of vegetable has common pests and diseases, especially viral diseases, downy mildew, soft rot, white spot, black spot, black rot, etc. In addition, crop rotation during production is needed.

(6) They are propagated by seeds, and both direct sowing or transplantation of seedlings are allowed.

Sub-context 1　Production of Chinese Cabbage

Chinese cabbage, also known as headed cabbage, yellow sprouts, stuffed cabbage, etc., is a variant in the genus *Brassica* of the Brassicaceae family. It is a 1–2-year-old herb with heads as edible organs. The bulbous leaves contain a lot of water, and the tender head contains nutrients such as carbohydrates, proteins, minerals, and vitamins. Chinese cabbage is used as autumn and winter vegetables in many regions of China while some are suitable for spring and summer cultivation.

I. Biological Characteristics

1. Botanical features

(1) Roots: Chinese cabbage is a plant with a shallow taproot system. Their taproots are relatively developed, thick at the top and thin at the bottom, with a large number of lateral roots. The taproots are not deep into the soil, generally about 60 cm long; lateral roots are mostly distributed in the soil layer 25 to 35 cm below the surface; and the root systems have a lateral expansion diameter of about 60 cm.

(2) Stems: Chinese cabbage has different-shaped stems at different development stages. The stems in the vegetative growth stage are called vegetative stems or shortened stems, and they grow into flower stems in the reproductive growth stage.

① Vegetative stems: From the seedling stage to the end of vegetative growth, vegetative stems grow mainly through leaves, with many leaves arranged closely in short internodes. Their vegetative stems are short, usually 4–7 cm, in a spherical or short conical shape, and called shortened stems.

② Flower stems: From the end of the rosette stage to the early heading stage, the top of the stems can develop into an inflorescence. Generally, their stems elongate and become flower stems during the late storage and reproductive growth stages. The top of the flower stem can turn into the main stem, and the buds between the leaf axils turn into lateral branches. Lateral branches can also have primary and secondary lateral branches, with longer base branches and shorter upper branches, giving the plant a conical shape. The height of the flower stems can reach 60–100 cm.

(3) Leaves: Chinese cabbage has 5 forms of leaves: cotyledons, primary leaves, rosette leaves, bulbous leaves, and flower stem leaves.

① Cotyledons: Chinese cabbage has two cotyledons, which are kidney-shaped, opposite, smooth, green, with petioles. When germination occurs, the cotyledons grow above the surface.

② Primary leaves: After the cotyledons are unfolded, the first pair of leaves that appear are primary leaves. Primary leaves are true leaves, which are opposite and arranged in a cross shape with the cotyledons, so this period is called the "crossing stage".

③ Rosette leaves: The leaves after the emergence of primary leaves and before the formation of bulbous leaves are called rosette leaves.

Rosette leaves have plate-like petioles with leaf wings. Rosette leaves are wide, wrinkled, and have wavy edges. They are composed of 2–3 leaf rings, arranged in 2/5 or 3/8 phyllotaxy. The main functions of rosette leaves are to perform photosynthesis and produce nutrients, so they are the main assimilation organ.

④ Bulbous leaves (head leaves): Bulbous leaves are metamorphic leaves, a storage organ for assimilation products, and a product organ. Bulbous leaves of Chinese cabbage are concentrically embraced to form a head. The numbers of bulbous leaves vary depending on the varieties, ranging from 30–80. The outer bulbous leaves turn green due to exposure to light, while the inner bulbous leaves turn white or light yellow. The bulbous leaves are mostly in a wrinkled and clasped state. The numbers and cohesion forms of bulbous leaves vary with different ecotypes, variants, and varieties. The bulbous leaves are alternate and embrace in three ways: folding, stacking, and twisting.

⑤ Flower stem leaves: The leaves attached to flower stems or flower branches are called stem leaves. As assimilation leaves that are grown during the reproductive growth period, flower stem leaves are alternate, and branching occurs between the leaf axils. Flower stem leaves are small and triangular in shape. The leaves grow along the stem, and become smaller and smaller as the growing part rises.

(4) Flowers: Chinese cabbage has compound racemes with 4 sepals, in green color. They have 4 petals, yellow or light yellow, arranged in a cross shape; 6 stamens, 4 strong and 2 weak, with nectaries at the base of filaments. 1 pistil, located in the center of the flower, with the ovary on the top. Chinese cabbage is a cross-pollinated crop. Their flowers are insect-pollinated flowers.

(5)Fruits: The fruits of Chinese cabbage are siliques with an elongated conical beak. Their seeds mature about 30 days after pollination. The pericarp of ripened fruits will crack vertically, and the seeds will fall off easily. The fruits are green at first and turn yellow when ripe.

(6)Seeds: The seeds of Chinese cabbage are spherical, reddish-brown or brown, with a few yellow. The thousand-seed weight is 2–3 g, with a seed life of 2–3 years.

2. Requirements for environmental conditions

(1) Temperature: Chinese cabbage likes a mild and cool climate. If the temperature during the growth period is higher than 25°C or lower than 10°C, the growth will be poor. The optimum temperature for seed germination is 20–

25°C, and the lowest germination temperature is 8–10°C. It takes a longer time to germinate under low temperatures. Germination is rapid under high temperatures of 26–30°C, but the seedlings are weak for excessive growth. Chinese cabbage in the seedling stage is highly adaptive to external temperature conditions. The optimum temperature range for Chinese cabbage seedlings is 22–25°C. While they can grow at higher temperatures of 26–28°C, their growth may be poor, and susceptible to viral diseases. The rosette stage requires strict temperature conditions. The optimum temperature of the rosette stage is 17–22°C. Too-high temperature will lead to too-fast growing but weak rosette leaves; Too-low temperature will retard the growth of rosette leaves. The heading stage has the strictest requirements on temperature conditions, and the optimum temperature range is 12–22°C. The temperature required in the early heading stage is slightly higher, and that required in the later heading stage is slightly lower. During the heading stage, the temperature difference between day and night should be increased. Generally, the optimum temperature difference between day and night is 8–12°C. The dormant period requires the lowest temperature conditions, preferably 0–2°C. Low temperatures below -2°C will lead to frost damage. The temperature of above 5°C will lead to vigorous respiration and excessive nutrient consumption. The optimum temperature for the regreening and bolting stages during the reproductive growth period is 12–22°C, and that for the flowering and fruiting stage is 17–22°C.

(2) Light: Chinese cabbage is a vegetable that requires medium lighting intensity. Reasonable dense planting is an important measure to obtain a high yield. However, too-weak light will seriously impact the growth and development of the plant. When the plants are too dense and the light is insufficient, the leaves will turn yellow and thin, and tend to grow upright, and the yield will be reduced greatly. Chinese cabbage is a long-day plant. Sunlight time requirements are not strict during its vegetative growth period, but shorter days are beneficial to the formation of heads. Its flowering and fruiting stages require longer sunlight.

(3) Water moisture: Chinese cabbage has a large leaf area and requires a lot of water. Chinese cabbage has shallow roots and an underdeveloped root system,

and cannot fully utilize the water in deep soil. Therefore, sufficient water supply is needed during its growth period. Dry soil during the seedling stage may easily lead to viral diseases caused by high temperatures and drought, so frequent watering is needed to lower the ground temperature and keep the soil moist. Watering appropriately during the rosette stage and abundantly during the heading stage to keep the soil moist can ensure the rapid growth of heads. Sudden dryness and sudden wetness should be avoided; otherwise, uneven watering will result in the cracking of heads. Watering quantity should be appropriately reduced during the late heading stage and harvesting stage to prevent the cracking of heads and facilitate harvesting and storage.

(4)Soil and nutrients: Chinese cabbage is a high-yield vegetable and has strict soil requirements. Sandy loamy soil, loam, and clay loam with deep soil, loose soil, and rich organic matter are preferred. It is suitable for cultivation in neutral, slightly acidic, or slightly alkaline soil.

Among the three elements of fertilizer, nitrogen and potassium fertilizers are needed the most. Nitrogen fertilizer promotes plant growth, and potassium fertilizer helps transport assimilated substances, thereby increasing yields. The appropriate combination of phosphorus and potassium fertilizers can improve disease resistance and quality. It is more sensitive to calcium contents, and insufficient calcium absorption can induce dryness, heartburn, and other diseases.

II. Types and Varieties

According to the growth period, Chinese cabbage can be divided into early-maturing varieties, medium-maturing varieties and late-maturing varieties. Early-maturing varieties generally have a growth period of around 50–60 days, medium-maturing varieties 70–80 days, and late-maturing varieties more than 90 days. According to physiological characteristics, it can be divided into spring Chinese cabbage, summer Chinese cabbage, and autumn and winter Chinese cabbage. Spring Chinese cabbage generally has a shorter growth period, with higher bolting resistance. Summer Chinese cabbage is characterized by good heat resistance. It can be divided into green type, white type, and green and white type according to leaf color. Generally, green varieties are more drought-resistant than white varieties, with less moisture, high dry matter content, and a longer storage period. Based on the features of heads, they can be further divided into 3 types. ① Numerous-leaf type: This type has many bulbous leaves but with lower unit leaf weight, with thinner middle ribs of leaves, and the ovate type mostly belongs to this type. ② Heavy-leaf type: This type has fewer but heavier bulbous leaves, with thicker middle ribs leaves. The flat-headed type mostly belongs to this type. ③ Intermediate type: This type is between the multifoliate type and the heavy-leaf type. According to the clinging methods of the bulbous leaves, they can be divided into stacking type, closing type, twisting type, etc.

According to the evolutionary process, head morphology, and ecological characteristics, Chinese cabbage subspecies can be divided into 4 variants: loose-leaf variant, semi-headed variant, flower-center variant, and headed variant. Moreover, the headed varieties are divided into 3 ecotypes: Ovate type, flat-headed type, and straight cylinder type, as shown in Fig. 4-2.

(1) Loose-leaf variant: It is a primitive type of Chinese cabbage, does not form heads, and has edible rosette leaves. It has strong stress resistance and poor quality, and has been gradually eliminated.

(2) Semi-headed variant: It has a loose head with an open head top and is in a semi-headed state. It has strong cold tolerance and no strict fertilizer and water requirements. Its rosette leaves and head leaves are edible products. It is mostly distributed in alpine regions such as Northeast and Northwest China. Its representative varieties include Shanxi Damaobian and so on.

(3) Flower-center variant: The plant is short and can form a solid head. The head top is not closed. The top of the bulbous leaves is white, light yellow, or yellow, and rolls outward to form the flower center. It is early-maturing, with a growth period of 60–80 days, and relatively heat-resistant, and can be used for early-maturing cultivation in autumn and spring cultivation. Its representative varieties include Beijing Fanxinbai, Ji’nan Xiaobaixin, Beijing Xiaozha 56, etc.

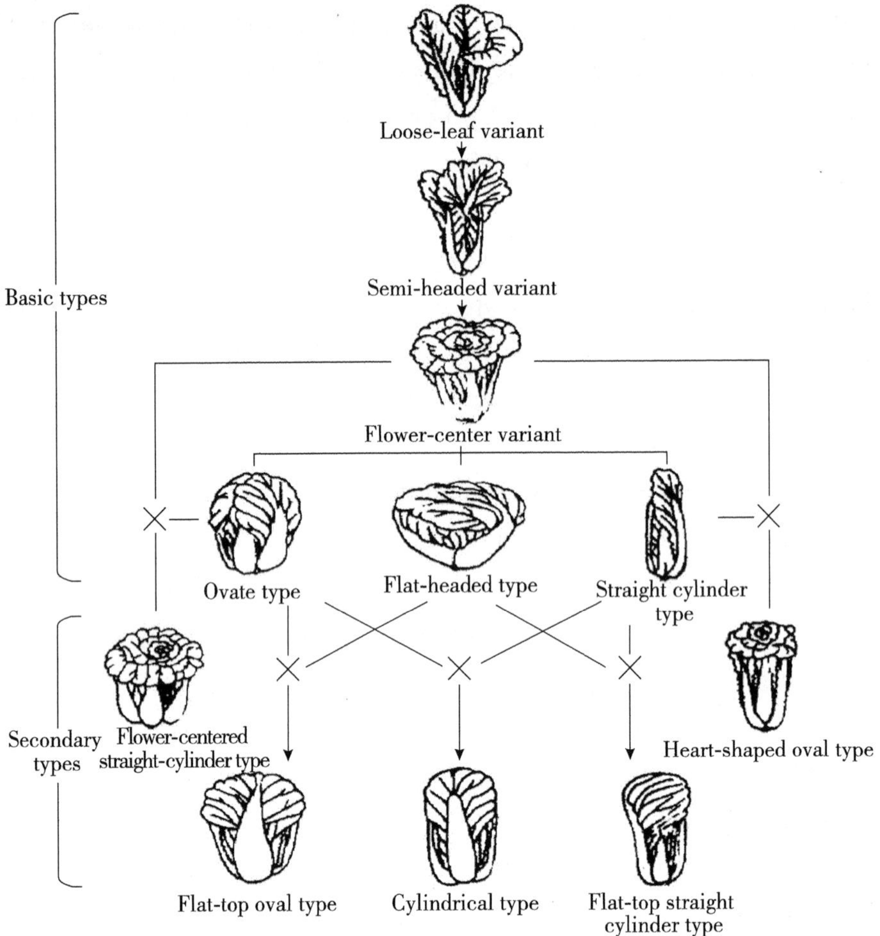

Fig. 4-2 Schematic Diagram of the Classification and Evolution Process of Chinese Cabbage (Li Jiawen, 1981)

(4) Headed variant: It is an advanced variant of Chinese cabbage and can form a firm head, with a nearly closed or completely closed top. It is of high quality, with high yields and storage resistance, and is the most commonly cultivated. This variant is further divided into the following 3 basic ecotypes:

① Ovate type (marine climate ecotype): Its head is ovate, with a sphericity index (i.e. the ratio of the head height to its cross-sectional diameter) of about 1.5. The head top may be round or obtuse, and the bulbous leaves are pleated, with numerous leaves and belong to the numerous-leaf type. Most varieties of the oval type have a growth period of 100–110 days, and a few early-maturing varieties have a growth period of 70–80 days. They are suitable for growing in an environment with mild climate, humid air, and moderate weather changes. Their cultivation center is in China's Shandong Peninsula, so it is called the marine climate ecotype. Its representative varieties include Shandong Fushan Baotou, Jiaoxian cabbage, etc.

② Flat-headed type (continental climate ecological type): This type has an obconical head, with a sphericity index close to 1. Its head top is flat and completely closed. The bulbous leaves are stacked and few in number, so they are classified as heavy-leaf type. Most varieties of the flat-headed type have a growth period of 100–120 days, and a few early-maturing varieties have a growth period of 70–80 days. It requires environmental conditions with a mild climate, a large day and night temperature difference, and sufficient light, so it is called a continental climate ecotype. However, it is adaptive to environments with drastic temperature changes and dry air to a certain extent. Its cultivation centers are in eastern Shandong and central and southern Henan. Its representative varieties include Shandong Guanxian Baotou, Heze Baotou, Henan Luoyang Baotou, and Shandong No. 4.

③ Straight cylinder type (cross-climate ecotype): This type has a slender and cylindrical head, with a sphericity index greater than 4. Its head top is pointed and nearly closed. Its bulbous leaves are twisted, with the habit of growing center leaves and bulbous leaves at the same time. Its growth period is 60–90 days. The cultivation centers of this type are in eastern Hebei and Tianjin. This type is a cross-climate ecotype of marine and continental climates, with strong climate adaptability. In addition, it has a habit of growing center leaves and bulbous leaves simultaneously, so it is easier to successfully induce it in various places. Its representative varieties include Tianjin Qingmaye, Hebei Yutian Baojian, Liaoning Hetou cabbage, etc.

III. Cultivation Season and Methods

In the past, Chinese cabbage was mainly cultivated in the open field in autumn and winter for consumption in autumn, winter, and spring. In recent years, with the adjustment of vegetable planting structure, more and more attention has been paid to off-season cultivation and perennial supply of Chinese cabbage. Therefore, the proportion of traditional autumn and winter cultivation has been declining yearly, while the proportion of spring and summer Chinese cabbage has been increasing. The cultivation methods of Chinese cabbage in different seasons are as follows.

1. Autumn and winter Chinese cabbage

Chinese cabbage likes cold and cool climates. Therefore, whether in South or North China, the main season of Chinese cabbage cultivation is chosen in the autumn and winter when the climate is mild and cool. Due to regional differences, in areas south of the Yangtze River, sowing can be done later in the South and earlier in the North. For example, sowing in Nanjing can be done in mid- to late August, in South China from September to November, and in Northeast China as early as mid- to late July. Considering that the optimum growth temperature of Chinese cabbage is in the range of 10–22°C, the three major diseases (downy mildew, virus disease, and soft rot) occur seriously at higher temperatures, and Chinese cabbage varieties generally have poor heat resistance in autumn and winter, if the growing season permits, the sowing date should be postponed as much as possible. In recent years, some excellent hybrids with short growth periods have been bred, making it possible to delay sowing.

2. Wintering Chinese cabbage

In Southwest and South China, such as Yunnan, Guizhou, Sichuan, Fujian, Guangdong, southern Guangxi, and Hainan, there is a tradition of cultivating wintering Chinese cabbage. The mild and cool climate conditions in winter can be utilized to sow seeds or culture seedlings from October to November in autumn and winter, and the finished products can be harvested and sold from March to April of the following year. However, this cultivation method still needs to pay attention to the problem of bolting in winter, especially in high mountains or plateau areas. On the one hand, varieties with strong wintering resistance and bolting resistance should be selected, such as Guizhou Huangdianxin No. 2, Youzhi No. 1, Fujian Lianjiangbai, etc. On the other hand, mulching films or small row covers can be adopted for covering and insulation, so as to improve quality and ensure output.

3. Spring Chinese cabbage

The cultivation of spring Chinese cabbage can not only adjust market varieties, but also increase profits for Chinese cabbage producers and sellers. In areas south of the Yangtze River and South China, spring Chinese cabbage is generally sown and cultured in plastic film row covers in mid-February and early and mid-March. If seedlings are cultured in common greenhouses or the greenhouses heated by heating wires, the seedling stage can be advanced. It can be cultivated under row covers from late February to late March, and harvested and marketed from late April to June depending on the field planting period. Early-maturing varieties with a short growth period, strong winter resistance, bolting resistance, and fast heading should be selected for cultivation, such as Shigatse No. 1, Huangdianxin No. 2, Luchunbai No. 1, No. 94–1, Beijing Xiaobaikou, and varieties introduced from Japan and South Korea like Qiangshi and Sijiwang.

4. Summer Chinese cabbage

The biological characteristics of Chinese cabbage determine that summer and autumn have always been the low season for the supply of Chinese cabbage. In recent years, with the progress in the selection of heat-resistant and disease-resistant varieties and summer facility cultivation technologies, it has become possible to cultivate summer Chinese cabbage in high-temperature areas in South China. The cultivation area continues to increase due to its good market and high prices. Summer Chinese cabbage can be sowed or cultured in summer from May to July, field planted from June to August, and harvested and marketed from August to September. Due to the hot weather in summer, with high temperatures, high humidity, and heavy rains, the varieties selected should be heat-resistant, disease-resistant, and early-maturing, with good heading capability and rapid growth, such as Xiaguan, Xiafeng, Xiaozha 55, Xiaozha 56, Xiayang, and Xiazhenbai. In terms of cultivation technology, it is advisable to adopt sun-shading facilities for seedling culture, as well as sun-shade and rain-proof cultivation or alpine cultivation.

IV. Cultivation Techniques

(1) Variety selection: When introducing and selecting Chinese cabbage

varieties, first, local dietary habits must be taken into consideration. Second, attention should be paid to local climate conditions, cultivation seasons, and crop matching, and varieties with a comparable growth period, disease resistance, high yields, and storage resistance should be selected. Third, avoid having monotonous varieties and brands, so it is best to plant 2–3 varieties every year and pay attention to variety matching.

(2) Soil preparation, ridging, and application of base fertilizer: Chinese cabbage has a deep and wide root system and a great growth volume. It is most suitable for loam and sandy loam soil with deep soil, rich organic matter, fertilizer retention, and good drainage and irrigation. Its required pH value is 6.5–7. Chinese cabbage should not be continuously cropped with cruciferous vegetables. Cucumbers, watermelons, allium, kidney beans, cowpeas, tomatoes, and rice are the most suitable previous crops of Chinese cabbage. Rational crop rotation is of great significance in reducing the spread of diseases.

① Soil preparation and ridging: Fine soil preparation and high ridges with deep furrows are important measures for Chinese cabbage to achieve uniform seedlings, full growth of seedlings, disease resistance, and high yields. During the soil preparation process, deep plowing and fine harrowing must be ensured to loosen the soil, level the field plots, and uniform fertility. It is required to have deep furrows and high ridges. Generally, the ridge width is 1.2–1.4 m, the furrow depth is 20–25 cm, the furrow width is 30 cm, and double-row planting is required. The ridge length should not exceed 30 m.

② Application of base fertilizer: Autumn and winter Chinese cabbage has a high yield and a long growth period, and requires a large amount of organic fertilizer as base fertilizer. Generally, 3,000–4,000 kg of decomposed manure, 25–30 kg of superphosphate, and 10 kg of potassium fertilizer are applied to each mu of field. 2/3 of the base fertilizer can be applied to the land and then plowed, and the remaining 1/3 can be applied to the furrows or holes right before ridging.

(3) Sowing and seedling culture: Chinese cabbage can be sown directly or transplanted after seedling culture. Direct sowing is characterized by a well-developed root system, strong drought adaptability, high disease resistance, and fast

plant growth, which is conducive to the completion of the heading process, will not damage the root system and leaves due to transplantation, and reduces the chance of soft rot. However, when the previous crop is harvested late or it is impossible for direct sowing in the open field due to persistent high temperature, drought, low temperature, and rainy weather (during spring sowing), seedling transplanting can be adopted to gain season.

Direct sowing can be divided into drill sowing and hole sowing (dibbling). Drill sowing is carried out on the ridge surface, with shallow trenches 0.6–1 cm deep at a certain row spacing. The seeds are sown evenly in the trenches, and the trenches are covered with soil. The seeds needed for each hectare is 2,250–3,000 g. Hole sowing (dibbling) is to dig shallow holes about 0.5 cm deep and 10–15 cm wide based on the predetermined row spacing and plant spacing. Each hole is sown with 10–15 seeds, and covered with soil to the extent that no seeds are visible, and pressed slightly to make the planting soil compactly connected. The seeds needed for each hectare is about 1,500 g. The field for direct sowing must have sufficient bottom moisture; otherwise drip watering sowing or post-sowing watering will be needed. The seeds will germinate at 18–22°C, and fully germinate in two days at optimum temperature. In order to prevent the drought and scorching sun in early autumn, sunshade nets may be covered above the seeds above the floating surface and removed till seedlings emerge. The direct sowing requires timely thinning of seedlings. The first thinning time is the crossing stage, and the second thinning time is at the 4–5 true leaf stage. Weak seedlings, inferior seedlings, and diseased seedlings must be removed (thinned) in time. The missing seedlings should be supplemented, and seedlings can be determined according to the predetermined spacing between plants in the resettling stage. Spring Chinese cabbage cultivated by direct sowing method should be first covered with mulching film and then with a row cover.

The seedling transplanting method uses nutrient soil blocks, nutrient pots, or plug seedlings. The culture soil is made up of decomposed manure and garden soil at a ratio of 1 ∶ (2–3), and then mixed with 0.1% compound fertilizer to serve as the seedbed soil. Level and rake the soil carefully before sowing, water it thoroughly, sow seeds, and cover the seeds with soil after sowing. In summer, it can be covered with a rain-proof shed and sun-shade net to prevent high temperatures and torrential rain, with timely watering. Generally, seedlings can be planted in fields after 15–18 days of culture. In recent years, 128-hole plug trays have been used to culture seedlings in many places. When culturing Chinese cabbage seedlings in spring, electric heating wires are often adopted to culture seedlings in greenhouses. The plug trays and nutrient pots are directly placed on the electric hotbed to culture seedlings. The ground temperature is controlled at 20°C, and the air temperature is controlled at above 15°C to prevent early bolting.

(4) Planting density: Reasonable dense planting is an important measure for high-quality and high-yield cultivation. Planting density varies depending on the varieties, climate, soil, fertilizer and water conditions, etc. The planting density for medium- and late-maturing plants should be 45,000–52,500 plants per hectare. The planting density of large varieties should be lower than that of medium- and early-maturing small varieties; that of rosette-leaf spreading varieties is lower than that of rosette-leaf varieties with upright leaves. According to the continuous testing results of the Yangtze River Basin Vegetable High Yield Collaboration Group, the planting density of late-maturing rosette-leaf spreading varieties should be 25,000–27,000 plants per hectare, while that for rosette-leaf varieties with upright leaves should be 31,500–34,500 plants per hectare. Among the early-maturing varieties, Xiafeng and Zaoshu No. 5, which have medium and smaller plant sizes, can have 52,500–75,000 plants per hectare, while Lubai No. 6, which has a slightly larger plant size, should have a planting density of 42,000–45,000 plants per hectare. The planting density of medium-maturing varieties is between those of early-and late-maturing varieties.

(5) Fertilizer and water management: According to the growth dynamics of Chinese cabbage and the laws of fertilizer and water requirements, while giving full play to the fertilizer effect, slow-acting base fertilizer should be used as the major fertilizer, to ensure that there is no shortage of fertilizer throughout the growth period. The growth volume in the seedling stage is small, with an underdeveloped root system and weak water and fertilizer absorption capability. However, autumn drought occurs frequently in South China, so it requires regular watering and a

small quantity of fertilizer to keep the soil moist. Moreover, seedlings should be thinned appropriately, and seedling lifting fertilizer should be applied once according to seedling conditions. The growth of the root system and leaves increases suddenly during the rosette stage, and 40% of the total topdressing should be applied in this period. It is advisable to apply a quick-acting fertilizer called Fa Ke fertilizer (fertilizer for plant growth). If fertilizer is deficient during this period, the rosette leaves will not grow well, and cannot be recovered even if makeup fertilizer is applied during the heading stage, so it is also called critical fertilizer. Generally, after the quantities of seedlings are determined, 7,500 –15,000 kg of decomposed manure or 150 kg of urea, 150 kg of calcium superphosphate, and 75–150 kg of potassium sulfate are applied per hectare in combination with intertillage and ridging, and then irrigate once. The principle is to let the soil dry and wet alternatively by alternating fertilizing and watering to prevent excess water from causing excessive growth.

The growth amount is the largest during the heading stage, which is the period when the fertilizer and water are absorbed the most. At this time, 50% of the total topdressing should be applied, and priorities should be given to the early and mid-heading stages, which are the so-called cylinder growing fertilizer and center growing fertilizer. Generally, decomposed manure should be applied by 22,500 kg per hectare as topdress immediately after the start of heading, together with 150 kg of urea, 150–225 kg of superphosphate and potassium sulfate, respectively. During the heading process of late-maturing varieties, depending on the growth conditions, compound fertilizer of about 375 kg per hectare should be topdressed as center growing fertilizer, and water should be irrigated heavily to keep the soil moist, and stopped till one week before harvest to avoid excessive moisture and intolerance to storage.

In recent years, due to the soil deterioration caused by the large-scale application of chemical fertilizers, various nutrient deficiencies have appeared, for example:

① Dry heartburn (heart rot). It is mainly caused by soil calcium deficiency or antagonism resulting from excessive application of chemical fertilizers (NH^{4+},

K^+, Mg^{2+}) despite no calcium deficiency, or by salinized soil resulted from drought or rainy weather with a salt concentration above 0.3%, all of which may lead to calcium absorption barriers of the root system. Its prevention methods include deep plowing and drainage, applying more organic fertilizers, preventing the excessive application of chemical fertilizers, planting selecting calcium deficiency-tolerant varieties, harvesting in advance appropriately, and applying 200-fold solution of calcium chloride twice during the initial heading stage, all of which has a certain effect.

② Sesame disease. Sesame disease refers to the appearance of many black sesame spots on the middle rib, which lowers the marketability. The causes of sesame disease are complicated and mainly include excess nitrogen, insufficient absorption of boron and iron during the heading stage, or excessive absorption of manganese, copper, and zinc in the heading stage. It can be prevented and controlled by applying more organic fertilizers, avoiding excessive use of chemical fertilizers, and planting sesame disease-resistant varieties.

③ Cracking disease. Cracking disease refers to longitudinal and transverse cracks occurring on the inner side of the middle rib, which will be followed by a corky state, and then by a browning and dry state; It is mainly due to boron deficiency. Boron deficiency is similar to calcium deficiency, and ammonium nitrogen and potassium can antagonize cracking disease. To avoid excessive application of nitrogen fertilizers, 12 kg of borax can be mixed with organic fertilizers and applied to one hectare of boron-deficient field. Boron-deficiency-tolerant varieties can be selected, and during the initial heading stage, twice spraying of 1 ∶ 200 boric acid to leave surfaces can have a certain effect.

(6) Prevention and control of diseases and pests. Soft rot, downy mildew, viral diseases, and anthracnose are the most serious hazards to Chinese cabbage in southern China, and comprehensive prevention and control measures should be adopted. For example, avoiding continuous cropping with cruciferous vegetables when crop rotation is implemented; choosing disease-resistant varieties; sowing seeds at the due time; covering seedlings with sunshade nets and insect-proof nets; covering them with silver-gray reflective mulching films; adopting deep furrows

and high ridges; applying more organic fertilizers deeply to avoid plant injury and chemical prevention and control, etc. During the seedling stage, aphids and viral diseases must be controlled. In the seedling and rosette stages, priority must be given to the prevention of downy mildew and anthracnose. 200-fold solution of 40% aluminum ethylphosphonate wettable powder, or 800-fold solution of 25% metalaxyl wettable powder, or 600–800-fold solution of 75% chlorothalonil powder can be used to spray once every 7–10 days for 2–3 times continuously. Soft rot can be prevented and controlled by spraying 200–250 mg/kg agricultural streptomycin every 7–10 days for 3–4 times in the early heading stage. Major pests include aphids, yellow-striped flea beetles, *Spodoptera litura*, cabbage caterpillars, diamondback moths, etc. They can be killed by pesticides such as BT emulsion, paclitaxel, 40% dimethoate emulsion, 25% deltamethrin, 20% cypermethrin emulsion, 50% aphid mist (pirimicarb), etc. which should be used alternately.

(7) Harvesting: Chinese cabbage can be harvested if the top of the head feels firm when pressed by hand. Early-maturing species must be harvested in time to prevent head cracking. For late-maturing species, it is useful to bundle the leaves before harvesting to prevent frost injury, or to tie up the outer leaves with straw 10–15 cm above from the top of the head. In South China, Chinese cabbage can be left in the fields and harvested by stages according to market needs. Chinese cabbage has a high yield, with early- and medium-maturing varieties yielding 45,000–52,500 kg per hectare and late-maturing varieties producing 75,000 kg of heads.

There are two methods of harvesting: cutting and pulling. Cutting refers to cut off the taproot with a knife or shovel; pulling means to pull the heads together with the taproots.

Post-harvest treatment: Cut Chinese cabbages with large wounds, so they should be dried in the sun after harvesting to dry and heal the wounds and reduce rotting losses during storage. The wounds of the pulled Chinese cabbage are small and easy to heal, but they can be put into the cellar only after the soil on the roots falls off. Before marketing, the outer old leaves and taproots should be removed, and the heads should be graded and packaged so that they can be sold clean.

Storage and preservation: Chinese cabbage has good storage tolerance and can be stored in piles (stacks), ditches, cellars, mechanical storage and air-conditioning storage.

「**Summary of Sub-context**」

This sub-context mainly introduces the general cultivation characteristics of cabbage vegetables and the cultivation technology of Chinese cabbage. Its focus is to understand the development process and cultivation techniques of Chinese cabbage.

「**Expanded Knowledge**」

Physiological Disorders of Chinese Cabbage

(1) Dry heartburn, also known as top burn, internal rot, internal browning, edge rot, etc., is a physiological disease caused by calcium deficiency.

(2) Cracked head and hollow head.

① Causes of cracked head: Untimely harvesting after head formation; Excessive water after encountering high temperature during the heading process; Excessive application of nitrogen fertilizer.

② Causes of hollow head: Potassium deficiency, calcium deficiency, and oxygen deficiency.

(3) No heading. Its causes may be improper sowing date (too early or too late), insufficient supply of fertilizer and water, early bolting, etc.

Sub-context 2 Production of Head Cabbage

Head cabbage is abbreviated as cabbage, also known as round cabbage, heading cabbage, etc. It is a variant of cabbage of *Brassica* cabbage species in the Brassicaceae family whose terminal buds or axillary buds can form a head (leaf ball). It is a biennial plant. The edible organ of head cabbage is its leaf head.

Head cabbage originated from the Mediterranean to the North Sea coast and was introduced to China in the 16th century. Head cabbage is characterized by high yields, strong adaptability, cold and heat resistance, rich nutrition, and storage and transportation resistance. It is commonly cultivated around the world and is one of the main vegetables in European and American countries. Head cabbage is cultivated all over China, and is the main vegetable in spring, summer, and autumn in North China, Northeast China, Northwest China, and other regions. It is also cultivated on a large scale in winter and spring in South China and other regions.

I. Biological Characteristics

1. Botanical features

(1) Roots: Head cabbage's taproot is thick, with a developed and dense root system. Its major root group is distributed in the soil layer within 60 cm deep, and most densely concentrated in the cultivated layer 30 cm deep. Its root system has a large lateral expansion, with a radius of up to 80 cm. Its root system has a strong regeneration ability and is suitable for transplanting.

(2) Stems: Head cabbage stems are divided into two types: shortened stems and flowering stems. During the vegetative growth period, the stems are shortened and hardly elongate, which are called shortened stems. However, the stems of Brussels sprouts are 50–100 cm high, the terminal buds are spread out, and the axillary buds can form many small heads. After entering the reproductive growth stage, the tops of the shortened stems of head cabbage will grow flower buds, which are called flower stems. The flower stems can branch and grow new leaves to form inflorescences.

(3) Leaves: Head cabbage of different growth stages can produce leaves with different shapes. Its cotyledons are embryonic organs, which are the first pair of leaves after emergence, kidney-shaped and opposite. The first pair of true leaves that grow after the cotyledons unfold are basal leaves, which are oval or elliptical opposite leaves vertical to the cotyledons, with long petioles and serrated leaf margins. During the period from the formation of basal leaves to the formation of heads, the leaves grow in a phyllotaxis of 2/5 or 3/8, and are attached to the

shortened stems in a rosette shape, thus called rosette leaves, which are powerful assimilation organs. After the formation of rosette leaves is the heading stage, when the mid-ribs of the newly grown leaves curve inward, wrap the terminal buds, and roll into a head (leaf bulb), thus the leaves are called bulbous leaves. The bulbous leaves have no petioles and are yellowish-white. The heads can be conical, spherical, or oblate in shape. After entering the reproductive growth stage, head cabbage grows flower stems, and the leaves on the flower stems are stem leaves, which are alternate. The leaves are smaller, with a pointed apex, a wide base, and no petiole or very short petiole.

(4) Flowers: Head cabbage has a compound raceme, with 3–4 levels of branches on the flower stems. The flowers are complete flowers, with 4 green sepals and 4 yellow petals arranged in a cross shape, 6 stamens, including 4 strong stamens, 4 nectaries, and 1 pistil. They are cross-pollinated and insect-pollinated flowers.

(5) Fruits. Head cabbage fruits are siliques, and their seeds mature about 40 days after pollination, with about 20 seeds in each pod of fruit.

(6) Seeds: Head cabbage seeds are spherical, reddish brown or dark brown, matte, with a thousand-seed weight of 3.3–4.5 g. The seeds generally have a service life of 2–3 years.

2. Requirements for environmental conditions

(1) Temperature: Cabbage is a cold-tolerant vegetable that likes mild and cool climates, but has strong adaptability. Its optimum temperature is 15–25°C for the growth period, 7–25°C for the rosette stage, and 15–20°C for the heading stage.

Cabbage has a strong adaptability to high temperatures. Its seedling stage and rosette stage can adapt to high temperatures of 25–30°C. Its high-temperature resistance decreases during the heading stage; high temperatures are adverse to heading. Drought weather will lead to loose heads, with declined quality. Cabbage is also very tolerant to low temperatures. Strong seedlings with 6–8 leaves can withstand low temperatures of -1– -2°C for a long period of time and -3– -5°C for a short period of time. Its heading stage can adapt to high temperatures. The heads of early-maturing varieties can withstand low temperatures of -3– -5°C for a short period of time, and the heads of medium- to late-maturing varieties can tolerate low temperatures of -5– -8°C for a short period of time. The bolting and flowering stages require higher temperatures, with the optimum temperature of 20– 25°C.

Head cabbage is a vernalized green vegetable, and its seedlings need to grow to a certain size before they can feel low temperature and complete vernalization. For example, early-maturing varieties should have 3 leaves and a stem diameter of more than 0.6 cm, and medium- to late-maturing varieties should have 6 leaves and a stem diameter of more than 0.8 cm. Its optimum temperature for vernalization is low temperature below 10°C, and vernalization is completed fastest at 2–5°C. Different head cabbage varieties have certain differences in their low-temperature requirements and required low-temperature time period to complete vernalization. Generally, early-maturing varieties require a wide range of high temperatures, while medium- and late-maturing varieties require the opposite. The low-temperature time required for vernalization is 30–40 days for early-maturing varieties, 40–60 days for medium-maturing varieties, and 60–90 days for late-maturing varieties. Within the appropriate temperature range, the lower the temperature, the shorter the time required for vernalization.

(2) Light: Cabbage likes medium-intensity light, and sufficient light during the vegetative growth period is beneficial to its growth. Most cabbage varieties do not have strict requirements on light intensity, so they can grow well both in southern areas with rainy weather and weak light, and in northern areas with strong light. These characteristics of cabbage are very beneficial to cultivation in protected areas in autumn, winter and spring. Cabbage is a long-day crop. After the vernalization stage, long-day conditions are conducive to bolting and flowering.

(3) Water moisture: Cabbage has large outer leaves and strong transpiration, but its root system is shallow, with weak absorption capacity, so it requires a humid environment. It grows well under the conditions of relative air humidity of 80%–90% and soil humidity (maximum field moisture capacity) of 70%–80%. If the soil moisture is low and the air is dry, the plant growth will be slow, the heading will be delayed, and the heads will be loose, seriously affecting the yield and quality. Cabbage cannot tolerate waterlogging; too much rain or poor drainage will brown

or kill the roots easily.

(4)Soil and nutrients: Cabbage is an adaptable vegetable; it does not have strict soil requirements. However, loamy soil, with fertile, loose soil, and strong water and fertilizer retention capacity is preferred. Its optimum soil pH range is 5.5–6.5. Acidic soils will lead to clubroots easily. Cabbage is resistant to salt and alkali. Cabbage is a fertilizer-loving and fertilizer-tolerant vegetable. It needs a lot of nitrogen fertilizer in the early stage, as well as potassium fertilizer and an appropriate phosphorus fertilizer in the heading stage. Throughout the growth period, the absorption ratio of nitrogen, phosphorus and potassium is 3 ∶ 1 ∶ 4. Calcium deficiency can easily lead to dry heartburn.

II. Types and Varieties

Head cabbage can be divided into common cabbage (with light green leaves), savoy cabbage (with wrinkled green leaves) and purple cabbage (with purplish-red leaves) based on the characteristics and color of the leaves, wherein common cabbage is the main variety cultivated in China. In terms of maturing period from sowing to the first harvest, common cabbage can be divided into early-maturing varieties (100–120 days), medium-maturing varieties (120–150 days), and late-maturing varieties(150 days or more). In terms of its shapes of head, it can be divided into the following three types.

(1) Pointed-headed type: This type has a small plant, an approximately conical head, a long central column, oval leaves, a thick middle rib, and low yield. It has strong winter resistance and is hard to bolt prematurely, so it is suitable for early-maturing cultivation in spring. Its representative varieties include Jixin (meaning chicken heart shape), Niuxin (meaning ox heart), etc.

(2) Flat-headed type: This type has a large plant, a large oblate head, and a shortened central column, and belongs to a medium-maturing or late-maturing variety. It is of high yields, good quality, and storage tolerance. It is commonly cultivated in the southern provinces of China, and its representative varieties include Black Leaf Xiaopingtou, Huangmiao, etc.

(3) Round-headed type: This type has a medium plant and a spherical head.

It is suitable for early- or medium-maturing cultivation. Its yield is not lower than that of the flat-headed type, but its bulbous leaves are crisp and tender and of good quality. It has weak winter resistance, and premature bolting is prone to occurring when cultivated as spring cabbage, so it is rarely cultivated. Its representative varieties include Jinzaosheng, Beijing Zaoshu, Chenghai Zaohua, etc.

In terms of the characteristics of leaves, common cabbage can be divided into:

(1) Common head cabbage: It has smooth leaves, no obvious wrinkles, slightly protruding middle ribs in the leaves, and green to dark green leaves.

(2) Purple head cabbage: Its leaf surface is the same as that of ordinary head cabbage, but its outer leaves and bulbous leaves are purplish-red, which turns to blackish-purple when fried, so it is not suitable for stir-frying, generally eaten cold and fresh, and is also cultivated as an ornamental.

(3) Wrinkled-leaf cabbage: Its color is similar to common cabbage. Its leaf surfaces are wrinkled due to developed and uneven mesophyll between the veins. Its bulbous leaves are soft and tender, with good taste, and can be stir-fried.

In terms of its length of growth cycle, head cabbage can be divided into:

(1) Early-maturing varieties: It takes 40–50 days from field planting to harvesting. Its representative varieties include Siji 39, Niuxin cabbage, Jixin cabbage, Zhonggan 12, etc.

(2) Medium-maturing varieties: It takes 55–80 days from field planting to harvesting. Its representative varieties include Zhonggan 15, Bigan No. 1, Yingchun, Xiyuan No. 4, etc.

(3) Late-maturing varieties: It takes more than 80 days from field planting to harvesting. Its representative varieties include Zhonggan No. 9, Huagan No. 2, Huangmiao, Heiye Xiaopingtou, etc.

III. Cultivation Season and Methods

Cabbage is highly adaptable, cold-resistant, and relatively heat-resistant. In northern China, it is mainly cultivated in open fields in spring, summer, and autumn. In recent years, facility cultivation of cabbage has been gradually developed, and a pattern of multi-crop planting and perennial supply has been formed.

In southern provinces of China, it can be produced in two or three seasons. It is sown from June to August and harvested in autumn and winter, which is called autumn cabbage or autumn and winter cabbage. The cabbage whose seeds are sown in mid- to late October and whose seedlings winter are harvested from April to June of the following year and thus called spring cabbage. The cabbage sown from March to May and cultivated in summer is called summer cabbage. Generally, autumn cultivation is most suitable. To sow and cultivate cabbage of different varieties with different maturing periods in rows can realize year-round supply of cabbage.

Different varieties of autumn and winter cabbage are sown and cultivated at different times, and have different harvest periods. The provinces in the middle and lower reaches of the Yangtze River mostly sow cabbage seeds in mid-July. If they are sown appropriately in June in advance and sunshade nets are utilized for culturing seedlings, good benefits can be achieved. They are generally harvested from late October to November. If sowing is delayed to late August to September, the harvest period will be from December to February of the following year. In Guangdong, early-maturing varieties are sown from September to October, which is more suitable than those sown from July to August.

The cultivation temperatures experienced by spring cabbage varieties are exactly opposite to those by autumn cabbage. The cultivation temperatures for autumn cabbage varieties gradually decline, so they can turn to reproductive growth after the smooth completion of vegetative growth, which can not only achieve the goal of harvesting the heads, but also complete the breeding of seeds. The overwintering cultivation of spring cabbage varieties has a long low-temperature period, which is adverse to growth but conducive to development. Therefore, its cultivation season is hard to control, and it is easy to cause premature bolting. The sowing period should be strictly controlled according to local climate conditions and local conditions. Generally speaking, in most areas of the middle and lower reaches of the Yangtze River, it is appropriate to sow seeds around mid-October. For example, if sown in Chongqing during Frost's Descent period, no premature bolting will occur, with good heading. Sowing and culturing seedlings in Xuzhou, Jiangsu in early and mid-November can also prevent premature bolting and realize heading. In northern Jiangsu, seedlings are sown in December. If mulching film or row cover is used for cultivation, the sowing date can be advanced appropriately.

Summer cabbage varieties are generally sown and cultured from March to May, and field-planted from May to June, and cultivated under sunshade nets. They are marketed in the off-season of vegetables from August to September. The hot climate during this period is not suitable for the growth of cabbage. Therefore, their yields are low, with low quality, high cost, and serious pests and diseases. In order to overcome these problems, the production can be done in suitable places, for example, cultivation in high mountains, to culture seedlings in row covers or open fields from March to April, field plant from April to May, and market from late June to early October. This method can help alleviate the supply shortage in autumn in South China.

IV. Cultivation Techniques

(1) Sowing and seedling culture: Autumn and winter cabbage grows slowly in the early stage, with strong root system regeneration capability, so it is suitable for seedling transplanting. June to July is the hot season in southern areas. Intense sunshine, high temperatures, and regular showers or heavy rains in this period bring difficulties to the culture of cabbage seedlings in autumn and winter. Therefore, measures to cool down and prevent heavy rain must be taken to ensure the culture of strong seedlings.

① Seedbed setting. The nursery site should be chosen at a plot with high and dry terrain, convenient drainage and irrigation, good ventilation, and loose and fertile soil, and the previous crop should be vegetable crops from different families. The land should be plowed and basked, and sufficient fully decomposed compost should be applied as base fertilizer, and no undecomposed organic fertilizers should be applied. The land should be prepared to make a ridge, usually a high ridge 1–1.3 m wide and 20 cm high, for sowing or transplanting seedlings.

② Cover setting. The principle is to set up a cover to prevent severe direct sunlight, allow appropriate penetration of sunlight, lower the temperature inside the cover, properly increase the relative humidity of the air in the shed, and create a microclimate suitable for the growth of seedlings.

Usually, row covers or greenhouses are used to culture seedlings under sunshade nets. When a greenhouse is covered, only the roof is covered but not the edges, and all the skirts around should be opened to facilitate ventilation. If a greenhouse with a double curtain trellis is adopted, the net can be fixed on the double curtain trellis to facilitate opening, closing, and uncovering. If conditions permit, the greenhouse can be covered with one net and one film; that is, to cover the top of the film with a sunshade net. In this way, the seedling growth rate can be increased by 20% compared with that covered by a single sunshade net. You can also cover the sunshade net with insect-proof net, to achieve a better seedling culture effect. When a row cover is covered with a sunshade net, a gap of 20–30 cm from the ground should be left around to allow ventilation.

③ Sowing. Both spread sowing and drill sowing are allowed. If drill sowing is adopted, furrows can be dug at a row spacing of 8–9 cm and a furrow depth of 1 cm, and be sown evenly. When the seedlings have about 3 true leaves, the seedlings can be separated once at a spacing of 8 cm × 8 cm. In order to reduce damage to seedling roots and let seedlings adapt to new environments as early as possible, it is advisable to use nutrient soil blocks, nutrient pots, plastic pots, and other seedling culture measures to protect the root system.

The plug seedling cultivation uses 132-hole plug trays and fills artificial substrates into the trays. The substrate can be peat, vermiculite, chaff ash, edible fungus culture leftovers, etc. Sow seeds in each hole and culture seedlings under the sunshade net, irrigate nutrient solution to grow the seedlings directly, thus realizing factory-style seedling culture.

④ Watering. In order to prevent the adverse effects of watering after sowing on the germination and emergence of seeds, the method of watering before sowing to ensure moisture for sowing can be used. After the emergence of seedlings, the watering volume should not be too large. For the newly emerged seedlings, water them once a day, and then once every other day. When the seedlings have 3 true leaves and are 5 cm tall, reduce the watering frequency. Watering in summer should be done when the weather, the ground, and the water are all cool. After rain, when the humidity of the seedbed is too high, dry soil can be spread to absorb moisture. When culturing seedlings under a cover, it is necessary to avoid excessive humidity in the bed, excessive growth of seedlings, or mildew root (damping-off).

⑤ Thinning. For dummy seedlings without transplantation, the seedlings are usually thinned out three times. The thinning is used to remove dense seedlings, weak seedlings, and low-quality seedlings. For the last thinning, one good seedling should be kept at a row spacing of 6–7 cm. For sowing in nutrient pots or nutrient soil blocks, leave 1 good seedling in each pot (block). The criteria for strong seedlings are flat cotyledons, symmetrical and unfolded base leaves, short internodes, sturdy stems, nearly round, and notchless base leaves, short petioles, and no diseases or insect pests. After the second and third thinning, apply plain manure in combination with watering to lift the seedlings, and then topdress the seedlings according to their growth conditions and needs.

(2) Land preparation and application of base fertilizer: Cabbage is afraid of waterlogging and requires good drainage, so it should be cultivated in narrow and high ridges. Generally, the width of a ridge (and a furrow) is 1.4–1.5 m, the furrow is 30–40 cm wide, and the ridge is 20–25 cm high. Among the three major elements of cabbage, the most absorbed one is potassium, the second is nitrogen, and the least absorbed one is phosphorus. 1,500–2,000 kg of decomposed organic fertilizers should be applied as base fertilizer for every mu of field.

(3) Timely field planting: When the cabbage seedlings have 7–8 true leaves, they should be field planted in time. Too-early field planting will lead to a long growth period, which is disadvantageous for management. Too-late field planting will lead to aging of seedlings, which is not conducive to high and stable yields. The suitable seedling age for field planting is about 40 days. The seedling age for planting should be shortened when the temperature is high and extended when the temperature is low. Planting distances vary depending on the varieties. Pointed and round-headed varieties can have wide rows of 60–70 cm, narrow rows of 40–50 cm,

and plant-to-plant spacing of 30–40 cm; Flat-headed varieties such as Jingfeng No. 1 can have wide rows of 80–90 cm, narrow rows of 60–70 cm, and plant-to-plant spacing of 50–60 cm.

(4) Field management: For autumn and winter cabbage, topdressing should be given priority to the prime growth period of rosette leaves and the early and middle heading stages. Topdressing should generally be performed 5 times. The first topdressing is to apply seedling lifting fertilizer when new roots appear after the field planting of seedlings, with relatively light dose, and generally thin and harmless fertilizers are applied. The second topdressing is done in the early growth stage of rosette leaves, and the amount and concentration of fertilizers should be increased, and generally dilute harmless manure and nitrogen fertilizer are used. The third topdressing is done in the peak growth period of the rosette leaves, when furrows are dug between the rows, organic fertilizers are mixed with nitrogen, phosphorus, and potassium chemical fertilizers, and all fertilizers are covered with soil and irrigated with water after the application. Topdressing should be done twice more in the early and middle heading stages, and no topdressing is needed in the late heading stage. In terms of the total dosage, nitrogen and potassium fertilizers should be the major ones, and phosphorus fertilizers should be applied appropriately. Generally, 1,500–2,000 kg of harmless manure and 50–75 kg of organic compound fertilizers are needed per mu.

Topdressing should be done together with water irrigation. Regular irrigation is required during the period after field planting when temperatures are high and rainfall is low. Although cabbage likes moisture, but dislikes waterlogging. The remaining water in the furrows must be drained immediately after each irrigation to prevent soaking roots from too long and subsequent rusty roots. After the heading process is completed, stop irrigation to prevent the leaf ball from cracking.

Intertilling and weeding should be done 2–3 times from field planting to the rime during the crop closure stage. The principle is to timely intertill and weed after heavy rain or irrigation, to prevent the soil surface from hardening or weeds from breeding. The soil must be hilled around the roots during intertilling and weeding.

(5) Prevention and control of pests and diseases: Cabbage pests, such as

cabbage caterpillars and *Spodoptera litura*, can do particularly serious harm, and almost always occur in the cabbage cultivation period. These pests can be controlled in the 2nd to 3rd instar stage of their larvae with 40% fenvalerate cream, 25% bifenthrin cream, or Chlorfluazuron (Yitaibao), flufenoxuron (Kasike), etc. The biopesticide BT emulsifier also has a good pest control effect. Aphids can also seriously harm overwintering cabbage seedlings when the weather is dry; they can be controlled by 40% dimethoate, 50% pirimicarb, etc. Diseases that harm cabbage include sclerotinia rot, soft rot, and black rot. Sclerotinia rot can be controlled by 40% wettable sclerotinia powder in the early onset. Black rot can occur during the entire growth period. Plants infected with black rot are prone to soft rot in the later stages. Its prevention and treatment methods include implementing crop rotation, which can prevent pests and invasion of germs. Moreover, infectious bacteria from flowing water should also be prevented. Black rot can also be controlled with 77% Kocide powder or 72% agricultural streptomycin sulfate wettable powder in the early onset.

(6) Harvesting: Generally speaking, the heads (leaf balls) can be harvested when they are compact to prevent heads from breaking or rotting due to rainwater or frostbite, which may affect their yield and quality. Generally, 3,000–4,000 kg of heads can be harvested every mu, or even 5,000–7,500 kg.

「**Summary of Sub-context**」

This sub-context mainly introduces the key points of cultivation of head cabbage. Its focus is on the key points of crop rotation arrangement and cultivation techniques of cabbage.

「**Expanded Knowledge**」

Which One is More Nutritious?

Common cabbage and purple cabbage are both variants of cabbage. In terms of nutritional content, there is not much difference between them. However, compared to common cabbage, purple cabbage is also rich in anthocyanins, a powerful antioxidant that can protect the human body from free radical damage and has anti-aging effects. It can also enhance blood vessel elasticity, inhibit inflammation and allergies, and improve joint flexibility. In addition, purple cabbage has a special aroma and flavor and can be boiled, stir-fried, served cold or made into pickles. Because it is rich in pigments, it is a good raw material for salads or Western food color matching. When stir-frying or cooking purple cabbage, a little white vinegar should be added before operation; otherwise it will turn dark purple after heating. Cabbage is suitable for stir-frying, boiling, etc. It can be made into soup with tomatoes or used as stuffing. Cabbage can inhibit cancer cells. Generally, cabbage planted in autumn has a higher inhibition rate, so you can eat more cabbage in autumn and winter. It is not advisable to buy too much since the nutritional content will decline after a few days.

Sub-context 3 Production of Cauliflower

Cauliflower, also known as broccoli and flowering vegetable, is a variant of the *Brassica* species in the Brassicaceae family, with curds (flower bulbs) as its products. Originating from the eastern coast of the Mediterranean, it was introduced to southern China in the mid-19th century and is now cultivated throughout China. Cauliflower is extremely nutritious and has a unique flavor, which plays an important role in enriching the colors and varieties of vegetables.

I. Biological Characteristics

1. Botanical features

(1) Roots: Cauliflower has strong root system, with developed fibrous roots. Its main root group is distributed in surface soil at a depth of 30 cm.

(2) Stems: During the vegetative growth period, the stems are slightly

shortened and the axillary buds on the stems are underdeveloped. After this period, the flower stems will grow.

(3) Leaves: Cauliflower rosette leaves are narrow and long, light bluish-green, with waxy powder. Its leaves naturally twist and roll inward to protect the curbs.

(4) Flowers: Cauliflower flowers are composed of flower stalks (axis), flower branches, and inflorescence primordia. They have inflorescence compound racemes, complete flowers, 4 strong stamens, and a superior ovary. They are cross-pollinated and insect-pollinated flowers.

(5) Fruits and seeds: Cauliflowers have siliques, with the capability of parthenocarpy. Its weight of one thousand seeds is 2.5–4.0 g.

2. Requirements for growth and development on environmental conditions

(1) Temperature: Cauliflower is a semi-cold-tolerant vegetable. It likes cool climates and avoids hot and drought conditions. Its normal adaptable temperature range is 13±7°C. Its optimum temperature for germination is about 25°C, its optimum temperature for growth in the seedling stage is 20–25°C, its optimum temperature for growth at the rosette stage is 15–20°C, and its optimum temperature for the curding stage is 17–18°C. Its curds grow slowly below 8°C, are susceptible to freezing below 0°C, and stop the growth and are easy to loosen above 24°C. Its optimum temperature for the flowering and podding stages is similar to that of the curding period. The temperature higher than 25°C or too low temperature will prejudice normal pollination and fertilization, resulting in empty pods. After germination, cauliflower seeds can pass through the vernalization stage at a temperature range of 5–20°C, with larger seedlings at 10–17°C passing through vernalization the fastest.

(2) Light: Cauliflower likes plenty of light and is also shade-tolerant. Strong light can easily turn the curds yellow and affect curd quality, so measures should be taken to shade the curds properly.

(3) Water moisture: Cauliflower requires a humid environment to grow and is less tolerant to drought and waterlogging. Sufficient water is required during the rosette stage and curd formation stage; otherwise, plant growth and curding will be affected. However, the water supply must be balanced, especially during the curding

stage. Too much water will easily result in loosened curds, moldy flower branches, and rusty roots.

(4) Soil nutrition: The suitable soil for cauliflower cultivation is loam or light sandy loam which is fertile and loose, with deep soil, convenient drainage, and strong water and fertilizer retention capabilities. The optimal soil pH value is 6.0–6.7. Cauliflower likes fertilizers and tolerates fertilizers. It needs more fertilizers throughout the growth period. It needs more nitrogen fertilizer in the early vegetative growth period, and balanced supplies of nitrogen, phosphorus, and potassium fertilizers should be ensured during the curding stage. Cauliflower is sensitive to micronutrients like boron, molybdenum, and magnesium. Boron deficiency will lead to shriveled growing points, curled leaf margins, hollow or cracked flower stems, rusty brown curds, and bitter taste; molybdenum deficiency will lead to deformed leaves, wine cup or whip shape curling, and slow growth; magnesium deficiency will result into yellowed leaves.

II. Types and Varieties

According to the length of the growth period, cauliflower can be divided into 3 categories: early-maturing varieties, medium-maturing varieties and late-maturing varieties.

1. Early-maturing varieties

It takes 40–60 days from field planting to starting to harvest the bulbs. The plant is smaller, more heat-resistant, and less wintery. When the stem thickness of the seedlings is 0.8 cm, they can undergo low temperatures to complete vernalization. Their curds are small and weigh 0.3–1.0 kg. Their main varieties include Xuefeng, Baifeng, Xiaxue 40, Xiaxue 50, Qilian Baixue, and Shenhua No.4.

2. Medium-maturing varieties

It takes 70–90 days from field planting to harvesting the curds. The plants of medium-maturing varieties are larger and more heat-resistant than early-maturing varieties, and have wide adaptability and strong winter resistance. When their seedling stems are 1.0 cm thick, they can accept low temperatures vernalization. Their curds are large, weighing more than 1.0 kg, with good quality and high yields.

Their main varieties include Shenhua No. 6, Xiahua 80, and Longfeng Teda 80.

3. Late-maturing varieties

Curds of late-maturing varieties should be harvested more than 90 days after field planting. Their plants are tall, intolerant to heat, and relatively cold-tolerant. They have strong winter resistance, and the seedlings can accept low temperatures vernalization only when their stems are more than 1.5 cm thick. Their curds are large, with a single curd weighing 1.5–2.0 kg or more, and the yield is high. Their main varieties include Shenhua No. 5, Shenlong Teda 180, Shenlong Teda 220, Shenlong Teda 240, Yidai Shenliang 120, etc.

According to the color of the curds, they can be divided into cauliflower and sprouting broccoli.

1. Cauliflower

They are white, and the very early-maturing varieties include Xiaxue 40, Holland Chunzao, etc. The medium-maturing varieties include Jinxue 88, Qilian Baixue, Fenghua 60, etc. The late-maturing varieties include Donghua 240, etc.

2. Sprouting broccoli

It includes two types: green cauliflower and broccoli. They are different from cauliflower in that: what is produced at the top of the main stem is not a curd (flower ball) composed of deformed flower branches, but a group of green, oblate spherical flower buds composed of completely normally differentiated flower buds.

III. Cultivation Season and Methods

Taking advantage of the different low-temperature requirements for flower bud differentiation of early-, medium-, and late-maturing varieties of cauliflower, sowing can be arranged in different seasons to basically ensure year-round supply.

Spring cauliflower in the middle and lower reaches of the Yangtze River is sown in sunward ridges from late November to early December, field-planted in the open fields from late February to early March, and harvested from early May to June. If mulching films and row covers are used for cultivation, the harvesting period can be advanced by about one month. Summer cauliflower is cultivated under sunshade nets from early to late June, field-planted from early to late July, and harvested from early September to early October. Some varieties can be harvested as early as late August, but with lower yield per unit area. Autumn cauliflower seedlings are cultured under sunshade nets from early July to early August, field-planted from early August to early September, and harvested from November to December. Winter cauliflower 120-day varieties are sown for seedling culture under sunshade nets from late June to early July. They are field-planted in open fields in early to mid-August and harvest from December to February of the following year; winter cauliflower 240-day varieties are sown in late July, field-planted from early to late September, and harvested from early April to early May.

In the cold northern areas, cauliflower can be cultivated in spring, summer, and autumn. Spring varieties are sown from February to March and harvested in June to July, summer varieties are sown in April and harvested in August, and autumn varieties are sown in June and harvested in September to October.

IV. Cultivation Techniques

1. Seedling culturing

The method of culturing cauliflower seedlings is basically the same as that of cabbage. Since the quantity of cauliflower seeds used is small, the seedling culturing techniques need to be more sophisticated.

Summer and early autumn are hot, with frequent showers. Therefore, seedbeds should be selected in a cool and ventilated place with slightly higher terrain, no scorching sun in the afternoon, or be provided with a shading shed on the seedbed to cool down and shield from rain. When culturing seedlings in autumn and winter, a warm and sheltered place can be used as a seedbed as the temperature gradually turns cool.

Cauliflower seeds are usually spread sowed, and it is better to sow seeds more sparsely in summer. The seedling stage is 30–60 days, and the seeding rate is 20–30 g/mu. During the seedling culturing period, attention should be paid to shading, cooling, and shielding from rain. Regular watering is needed when dry, and temporary planting is needed when the seedlings have 3–4 true leaves. Under suitable conditions, seedlings will emerge 2–3 days after sowing, and can be field-planted when they have 5–6 true leaves.

2. Field planting

Although cauliflower likes a humid environment, it has poor waterlogging tolerance. Therefore, in rainy areas and places with high groundwater levels, deep trenches and high borders should be used to facilitate drainage. This is a key to the success of cauliflower cultivation.

Cauliflower has stricter soil nutritional requirements than head cabbage. Cauliflower should be cultivated in loam to clay loam, and sufficient base fertilizer should be applied. Early-maturing cauliflower varieties have a short growth period, and the base fertilizer used should be mainly quick-acting nitrogen fertilizer. For medium- to late-maturing cauliflower varieties, the base fertilizer should be manure combined with phosphorus and potassium fertilizers. 2,500–5,000 kg of manure or 300 kg of vegetable-specific compound fertilizer should be applied per mu. Cauliflower is sensitive to micronutrients such as boron and molybdenum. About 50 g of borax or boric acid and about 10 g of ammonium molybdate or sodium molybdate can be mixed and made into an aqueous solution per mu, and applied to the field planting holes.

The nutritional areas for field planting of cauliflower vary depending on the varieties. Early-maturing varieties are usually planted in 3 rows in a 1.7 m wide ridge, and arranged in a triangle; or planted in 2 rows in a 1.3 m wide ridge, with a plant spacing of 30–40 cm, and 2,500–3,000 plants per mu of field; medium-maturing varieties are generally planted in 2 rows in a 1.3 m wide ridge, with a plant spacing of 40–50 cm, and 2,000–2,300 plants per mu of field. For late-maturing varieties, the plant-to-plant spacing should be 50–60 cm, and 1,600–1,800 plants should be planted per mu.

3. Field management

Field management work of cauliflower includes topdressing, irrigation, intertillage, weeding, leaf bunching, etc.

Cauliflower should have different water and fertilizer management due to different varieties and cultivation periods. Early-maturing cauliflower varieties have a short growth period and require urgent water and fertilizer, so quick-acting fertilizers should be applied regularly in stages. For medium-maturing cauliflower varieties, quick-acting fertilizers should also be applied regularly in stages during the leaf cluster growth period; during the curding period, the temperature is suitable for growth and development, and the applied amount of fertilizer should be increased to promote the growth of leaves and curds. For spring cauliflower that overwinters in the middle and lower reaches of the Yangtze River, during the leaf cluster growth period, attention should be paid to water and fertilizer management according to climate conditions to promote leaf cluster growth and lay a solid foundation for early maturation and high yields of the cauliflower. When the cauliflower is about to curd, the leaves on the layer adjacent to the curd are lighter in color or have obvious wax powder. This is a sign of curding, and quick-acting fertilizer should be applied 1–2 times as soon as possible. Cauliflower should be mainly applied with nitrogen fertilizer at all growth stages. When it enters the curding stage, phosphorus and potassium fertilizers should also be applied appropriately. Generally, for early-maturing varieties, 1,300–2,000 kg of decomposed manure or 50 kg of nitrogen, phosphorus, and potassium compound fertilizers should be applied per mu. For medium- to late-maturing varieties, the quantity of potassium fertilizer and nitrogen fertilizer should be appropriately increased on the basis of the fertilizers applied to early-maturing varieties. Boron and molybdenum play an important role in the curding process. During the growth period of the plant, 0.1%–0.2% borax and ammonium molybdate spray can be used for extraradical topdressing.

Cauliflower likes moisture and requires a lot of water during its entire growth process. The period of vigorous growth of leaf clusters and curds is a water-critical period. If its water demand cannot be met in time, the size of curds will often be affected. Therefore, in the case of high temperatures and prolonged drought, it is necessary to irrigate in time and remove the remaining water promptly to avoid soaking for too long and causing root rotting. Spring cauliflower generally does not need irrigation. In areas with a lot of rain, drainage must be strengthened to prevent water stains.

The work of intertillage, weeding, soil hilling, pest and disease prevention, and control of cauliflower are the same as those for cabbage.

Bunching leaves is one of the techniques to ensure the quality of cauliflower.

It is usually done when the curds emerge. The method of bundling leaves is to wrap the curds with a few large leaves outside the curd, and then gently tie them in a circle with straw or other materials. Be careful not to damage the leaves when tying. This method is often practiced in areas with obvious frost. In South China, the method of breaking off the large leaves close to the curds and covering them on the curds is often used.

4. Harvesting

The number of days from emergence of curds to harvesting varies with variety and climate. Early-maturing varieties form curds quickly when the temperature is relatively high and can be harvested in about 20 days. Medium- to late-maturing varieties often take about 1 month in late autumn and winter for curds, while late-maturing varieties take more than 20 days from the emergence of curds to harvesting in spring.

The curding of cauliflower is sometimes very inconsistent between individual plants and should be harvested in stages. The standard for harvesting is that the curds are fully grown, have a round surface and the edges have not yet spread out. When harvesting, several leaves under the curds can be cut off at the same time to protect the curds and facilitate packaging and transportation. The yield of cauliflower varies depending on the varieties, cultivation seasons, and management levels. Generally, the yield of early-maturing varieties is 750–1,500 kg per mu. The medium- to late-maturing varieties have a yield of about 2,500 kg, and the high-yielding varieties may have a yield of more than 4,000 kg.

V. Common Cultivation Issues and Their Prevention Strategies

In cauliflower production, we sometimes encounter early flowers, blue flowers, hairy flowers, purple flowers, and hollow stems, which seriously affect the quality of cauliflower. So attention should be paid to solving the above problems. Early flowering means the early formation of small balls. Sometimes, cyan or hair-like objects, called blue and hairy flowers, appear on the surface of the curd. Sometimes the surface of the curd has an uneven purple color, which is called purple flower.

The main reason for early flowering in cauliflower is that the plant encounters low temperatures prematurely when it is still young, which induces premature differentiation and formation of the curd. Early-maturing varieties may be caused by sowing too late, or not paying attention to selecting varieties according to the number of days during the seed-saving process, resulting in a mixed harvest of seeds with strong and weak winter resistance. Therefore, it is necessary to understand the characteristics of the varieties and pay attention to the sowing period. At the same time, field management should be strengthened, and water and fertilizer should be supplied in a timely manner so that plants can complete vegetative growth within a certain growth period.

The reason for the appearance of hairy flowers on cauliflower may be that the exposed tops of the flower branches of the cauliflower stretch irregularly due to the influence of climate. This phenomenon often occurs in field cultivation. Whether the main cause is climatic factors or the role of recessive genes is unclear. Mastering the appropriate sowing period according to the characteristics of the variety can prevent this phenomenon from happening.

Blue flowers are the result of the curd producing green bracts, sepals, etc. and even small flower buds, which is the result of non-sequential and sudden flower bud differentiation. It may be caused by the low temperature and foggy environment during the development of the curd. Therefore, the sowing period should be adjusted well in areas with foggy winters. The purple flowers are caused by the sudden low temperature encountered after the curd is formed, and the glycosides in the curd tissue are converted into anthocyanins. This is also related to certain breed characteristics.

Hollow stems are caused by nutrient deficiency, especially boron deficiency during the formation of curds. Symptoms of nutrient deficiency will appear on the stems, leaves, and bulbs. The curds will begin to appear water-soaked, and the surface will become discolored. Finally, the stem will become hollow and the cavity wall will become discolored. If it continues to develop, the flower buds and flower branches inside the curd will rot, and the curd will also become hollow. The control method is to spray 0.1%–0.2% boric acid on the leaves.

「**Summary of Sub-context**」

This sub-context mainly introduces the key points of cauliflower cultivation. Focus on mastering the key points of cauliflower crop arrangement and cultivation techniques.

「**Expanded Knowledge**」

Colorful Cauliflower

Colorful cauliflower is a herbaceous plant of the Brassicaceae family, including purple cauliflower, yellow cauliflower, light green cauliflower, etc., which are varieties selected from cauliflower. The mature cauliflower curds generally have a transverse diameter of 20–30 cm and a diameter of 10–20 cm. They are composed of a plump main axis and numerous fleshy flower stalks. In addition to being cultivated as ordinary cauliflower for commercial sale, colored cauliflower can also be grown in pots for small family vegetable gardens or ornamental cultivation, with promising prospects.

Yellow, purple, and pagoda-shaped colorful cauliflowers are different from conventional white cauliflower and green cauliflower, adding a beautiful scenery to the table. It not only has high edible value, but also has richer nutrition than white cauliflower. Each 100 g of fresh balls contains 2.4% protein, 3% carbohydrate, and 88 mg of vitamin C.

Sub-context 4 Production of Chinese Kale

Chinese kale, also known as Chinese broccoli or *Brassica alboglabra* Bailey, is a biennial herbaceous plant belonging to the Brassicaceae family. It is cultivated for its tender flower stalks and young leaves. Chinese kale is a specialty vegetable in China, known for its rich nutritional content, crisp texture, and unique flavor. It originated in the southern regions of China and is mainly found in Guangdong,

Guangxi, Fujian, and Taiwan. There are also limited cultivation efforts in cities like Beijing, Shanghai, Nanjing, Jinan, and Hangzhou.

I. Biological Characteristics

1. Botanical characteristics

Chinese kale is a shallow-rooted vegetable with strong regenerative abilities in its roots, primarily distributed within the 15–20 cm soil layer. Typically, the plant grows to a height of 40–50 cm with a spread of 35–45 cm. The stems are green, upright, and slightly shortened. The newly formed flowering stems are fleshy, and the internodes are relatively sparse and green. These young flowering stems, known as tsai-tai in Chinese, are the edible organs of the plant. The leaves are single, alternate, and densely green in color, with a waxy surface. They are ovate, elliptical, or nearly circular in shape, and can be smooth or slightly wrinkled, with green petioles. Chinese kale produces compound racemes, bearing complete flowers that are either white or yellow in color. They are cross-pollinated, and insect-pollinated flowers. It forms siliques containing nearly round seeds that are brown or dark brown in color, weighing approximately 3.5–4.0 g per thousand seeds. The growth cycle of Chinese kale includes the germination stage, seedling stage, leaf cluster growth stage, tsai-tai formation stage, and flowering and fruiting stage.

2. Requirements for growth and development on environmental conditions

The optimal temperature for Chinese kale seed germination and seedling growth is between 25–30°C. For the growth of leaf clusters and tsai-tai formation, the suitable temperature range is 15–25°C. A significant temperature difference between day and night is favorable for the formation of tsai-tai. Temperatures above 30°C or below 15°C both adversely affect the development of tsai-tai. Early- and medium-maturing varieties can rapidly differentiate flower buds at high temperatures around 27–28°C, and low temperatures do not significantly promote floral bud differentiation. In contrast, late-maturing varieties benefit from lower temperatures and longer periods of low temperatures to facilitate flower bud differentiation. Chinese kale is a long-day plant, but it does not have strict requirements for the length of daylight. Sufficient sunlight is necessary for the growth of leaf clusters and the formation of flower stalks. Chinese kale is best cultivated in loose, fertile, well-drained sandy loam soil or loam. It is tolerant to fertilizers, prefers moisture, dislikes drought, and is sensitive to waterlogging. Throughout its growth stages, the plant absorbs potassium the most, followed by nitrogen, with phosphorus being the least absorbed nutrient.

II. Types and Varieties

Chinese kale can be categorized into white-flowered Chinese kale and yellow-flowered Chinese kale. Yellow-flowered Chinese kale is less commonly cultivated, whereas white-flowered Chinese kale has a larger cultivation area and wider distribution. It can be classified into the following three types:

(1) Early-maturing varieties: These varieties are relatively heat-tolerant and have high yields. They are usually sown in the summer and fall in the southern regions of China, with harvests occurring from September to December. Representative varieties include fine-leaf early Chinese kale, wrinkled-leaf early Chinese kale, and willow-leaf early Chinese kale.

(2) Medium-maturing varieties: These varieties are less heat-tolerant than early-maturing ones and less cold-resistant than late-maturing varieties. In the southern regions of China, they are typically sown from September to November and harvested from November to February of the following year. They can also be cultivated in late spring. Representative varieties include Hetang Chinese kale, Dengfeng medium-maturing Chinese kale, Fujian Chinese kale, and Taiwan mid-flowering Chinese kale.

(3) Late-maturing varieties: These varieties are not heat-tolerant, but they are relatively cold-resistant and should not be sown too early. They are suitable for winter cultivation. Late-maturing varieties are usually sown from October to December and harvested from January to April of the following year. Representative varieties include Copper Chinese kale, wrinkled-leaf late Chinese kale, and Sanyuanli late Chinese kale.

III. Cultivation Techniques

Chinese kale is usually propagated through seedling transplantation, although direct sowing is also an option. During the summer, it is essential to provide shade and protection against rain for seedlings, while in winter, insulation is crucial. Transplanting is necessary when the seedlings have 5–6 true leaves and are 20–30 days old. Prior to transplanting, it is crucial to apply sufficient basal fertilizer. For early-maturing varieties, the planting density is 15–20 cm apart in both rows and columns, while for medium- to late-maturing varieties, it is 20–25 cm apart. It is necessary to water the seedlings slowly 3–5 days after transplanting, perform intertillage, and apply urea or other readily available nitrogen fertilizers along with watering every week after the slow seedling stage. When the plants have flower buds, it is required to increase the amount of readily available nitrogen fertilizers appropriately and supplement them with compound fertilizers. After harvesting the main stalks, it is crucial to apply substantial topdressing and water to facilitate the growth of side flower stalks. During the leaf growth period, the soil moisture needs to be maintained at a moderate level, and during the flower stalk formation period, the relative humidity in the soil should be kept around 80%–85%. It is essential to ensure timely drainage and prevent waterlogging after rainfall.

The ideal harvest time for Chinese kale is when the main flower stalk is at the same height as the outer leaves and the initial flowers start to appear. During the main flower stalk harvest, it is recommended to leave 4–5 basal leaves intact and cut the tsai-tai, allowing the side shoots to grow. Side shoots can be harvested approximately 20 days later. When harvesting side shoots, it is crucial to leave 1–2 basal leaves on the plant to encourage the formation of new side shoots.

「**Summary of Sub-context**」

This sub-context focuses on mastering the cultivation technology and the field management technology of Chinese kale.

「**Expanded Knowledge**」

Medicinal Value of Chinese Kale

(1) Chinese kale has therapeutic effects on excessive heat in the gastrointestinal system, insomnia caused by staying up late, and gum swelling and bleeding due to a lack of vitamin C. It is popularly prepared as Chinese kale tea, where sliced Chinese kale is boiled to make a clear soup, which is then consumed warm. In severe cases of gum bleeding, watercress and lotus root can be added to the preparation.

(2) Chinese kale is rich in vitamins A and C, calcium, proteins, fats, and plant carbohydrates. It has the ability to moisten the intestines, reduce internal heat, and stop gum bleeding.

(3) It is essential to consume Chinese kale in moderation. The quantity should not be excessive, and the frequency of consumption should not be too high. Traditional Chinese medicine believes that excessive consumption of Chinese kale can deplete vital energy. Prolonged consumption may inhibit sex hormone secretion. The traditional Chinese medicine text *Ben Cao Qiu Yuan* (Seeking the Origin of Materia Medica) has documented that Chinese kale is "sweet, pungent, and cold, consuming qi and damaging blood".

「**Reviewing and Thinking Questions**」

I. Explanation of Terms

1. Crossing.
2. Head index.
3. Critical fertilization.

II. Gap Filling

1. The leaf-curling methods of heading variation in Chinese cabbage are ________ ,________ , and________ .

2. Heading Chinese cabbage is classified into________, ________, and ecological types.

3. The three major diseases of Chinese cabbage are ________, ________, and ________.

4. Cauliflower needs its leaves folded over the flower head when it reaches a diameter of ________ cm. This is done to________.

5. Early flowering and florets in cauliflower are primarily due to________, ________, and other factors.

6. Common cabbage can be categorized based on head shape into________, ________, and ________ varieties. ________ are often early-maturing and suitable for spring cultivation.

7. Heading cabbage is a vernalized________vegetable. It generally requires a temperature of________°C to complete vernalization fastest. The lower the temperature, the ________ time needed for vernalization within the suitable temperature range.

III. Choice Questions

1. Cauliflower has special requirements for mineral elements. When lacking which element, it often causes hollow or cracked stems, rusty-brown color in the curd, and a bitter taste? (　　)

A. N　　B. K　　C. B　　D. P

2. Excessive use of fertilizer during the heading period that leads to non-heading is due to which nutrient deficiency? (　　)

A. N　　B. P　　C. K　　D. Ca

3. What is the cause of internal tipburn in Chinese cabbage? (　　)

A. Ca deficiency　　B. Mg deficiency

C. B deficiency　　D. Mo deficiency

IV. Thinking and Answering

1. Briefly describe the high-yield and high-quality cultivation techniques for autumn Chinese cabbage.

2. Briefly explain the key cultivation techniques for Chinese cabbage during spring and summer.

3. How can premature bolting in spring cabbage be prevented?

4. How can a year-round supply of heading cabbage be achieved?

5. Summarize the characteristics of water and fertilizer management in leafy vegetables like Chinese cabbage.

6. Analyze the primary issues in stem mustard production and suggest countermeasures.

7. Compare the similarities and differences in the cultivation techniques between cauliflower and heading cabbage.

8. How can the occurrence of florets, hairy flowers, green flowers, blue-and-white flowers, purple flowers, and hollow stems be prevented in cauliflower?

9. Provide an overview of cultivation techniques for Chinese kale.

Learning Context 5　Production of Root Vegetables

「Learning Objectives in This Context」

Master the main types and common characteristics of root vegetables, understand their biological features and basic knowledge related to root vegetable production, learn to select appropriate root vegetable varieties based on cultivation facilities and seasons, develop skills in planning root vegetable production, and execute the production process accurately.

「Analysis of Tasks in This Context」

Selecting and breeding varieties, preparing land, planting and adjusting plants, managing fertilization and irrigation, preventing and controlling pests and diseases, and post-harvest handling.

「Introduction」

Root vegetables are essential crops in the realm of vegetables. Root radishes, in particular, are nutritionally rich and widely enjoyed by people across the country. Burdock is a significant vegetable for export, contributing significantly to foreign exchange earnings.

Overview

I. Classification of Root Vegetables

Root vegetables refer to vegetables whose enlarged, fleshy roots are consumed as food. This category includes radishes, mustards (common root mustard, commonly referred to as "kohlrabies"), horseradish, kudzu, carrots, kohlrabies, turnips, rutabagas, sugar beets, root celeries, parsnips, burdock, *Jerusalem artichoke*, *Tragopogon pratensis*, etc. (Fig. 5-1). Currently, in southern China, radishes and carrots are the main root vegetables cultivated. Most of the others are often used in specialized cuisines. Radishes are believed to have originated from wild species along the Mediterranean coast and warm coastal areas of Southeast Asia. They are one of the oldest cultivated plants. Carrots originated in the foothills of the mountains in Afghanistan. They were introduced to Europe via Iran and Türkiye between the 9th and 10th centuries and later reached China in the 13th century through Iran. From China, carrots were further introduced to Japan. Carrots are one of the most widely distributed root vegetables globally.

Root vegetables prefer a mild and cool climate during their growth period. In the first year of growth, they develop fleshy roots with well-developed parenchyma cells, which are suitable for storing a large amount of nutrients, including vitamins, minerals, sugars, and starch. In the second year, they bolt, flower, and set seeds. They generally undergo vernalization in low temperatures and photoperiodic induction under long daylight conditions. They thrive in loose and deep soil, and are propagated through seeds. Root vegetables are resilient for storage and transportation and can be consumed fresh, stir-fried, pickled, and preserved. They are among the favorite vegetables of many people. Some premium varieties and processed products, such as Yunnan yellow mustard greens, Yunnan shredded radishes, and Chongqing Peiling pickled vegetables, are also essential export commodities.

Rich in carbohydrates, vitamin C, and minerals, root vegetables play a crucial role in regulating physiological functions. Among them, carrots hold the highest nutritional value. These vegetables primarily develop their edible parts in the form of root systems and thrive in deep, fertile, loose, and well-drained sandy loam soil. Cultivating them in soil that is poor, sticky, sandy, or gravelly can lead to the formation of deformed roots, affecting the overall quality. Root vegetables are plants that undergo cross-pollination.

Fig. 5-1 Different Root Vegetables

II. Characteristics of Root Vegetables

Among root vegetables, radishes and carrots are the most commonly cultivated. Root vegetables are primarily native to temperate regions and are categorized as cold-resistant or semi-cold-resistant annual or biennial vegetables. They undergo vernalization in low temperatures and photoperiodic induction under long daylight conditions. Root vegetables are highly nutritious. Radishes have the effects of clearing heat, detoxifying, moistening the lungs, relieving coughs, promoting bowel movements, facilitating urination, and aiding digestion. Burdock is effective in promoting blood circulation, preventing strokes, kidney diseases, hemorrhoids, constipation, and lowering blood sugar. Carrots are rich in various carotenes, which can prevent night blindness, respiratory diseases, and enhance resistance to pathogenic infections, and aid digestion; in addition, they have positive effects in treating indigestion, chronic diarrhea, coughs, and preventing cardiovascular diseases, and possessing anti-cancer properties. Various types of root vegetables have been used in traditional Chinese medicine for their diuretic and nerve-stimulating effects. Additionally, root vegetables play a significant role in traditional Chinese pickling processes. Furthermore, they provide nutrient-rich, fresh, and succulent green fodder for the livestock industry.

Root vegetables have a long history of cultivation in China, leading to the development of several excellent varieties. Examples include Xinlimei from Beijing and Weixian radish from Shandong.

Root vegetables share several common characteristics in cultivation: ① They are deep-rooted plants, and their fleshy roots are the edible parts. ② They thrive in deep, fertile, loose, and well-drained sandy loam soil. ③ They are usually sown directly from seeds and are not transplanting-friendly. ④ Most of them are cold-resistant or semi-cold-resistant biennial plants. ⑤ They undergo photoperiodic induction under long daylight conditions. ⑥ They are cross-pollinated plants, requiring strict isolation during seed collection. ⑦ Root vegetables within the same botanical family often face common pest and disease problems. Therefore, continuous cropping of these vegetables is not advisable.

Sub-context 1　Production of Radish

Radish is an annual or biennial herbaceous plant belonging to the *Raphanus* genus of the Brassicaceae family. It forms enlarged fleshy roots and is also referred to as Laifu or Lufu in Chinese. Originally native to China, radish has a long history of cultivation and exhibit strong adaptability, leading to widespread cultivation across the world. In European and American countries, small-sized radishes are predominantly cultivated, while in Asian countries, large-sized radishes are the primary focus. In northern regions of China, the cultivation area of autumn radishes is second only to Chinese cabbage, making it one of the major vegetables grown in the autumn and winter seasons. The fleshy roots of radishes are nutritionally rich and contain amylase, aiding in digestion when consumed raw. Moreover, these roots contain antimicrobial substances, which contribute to their medicinal value by helping dispel phlegm, stopping diarrhea, and promoting diuresis. Radishes can be consumed raw as a fruit, pickled to make sour radish strips or diced radishes, dried to create radish flakes, or used as decorative carving materials in banquets and feasts.

I. Biological Characteristics

1. Botanical features

(1) Roots: Root vegetables are deep-rooted plants and are best cultivated in deep soil, rich in organic matter, has good fertility and water retention properties, and allows for easy drainage.

The fleshy taproot of root vegetables can be divided into three main parts (Fig. 5-2):

① Root head (top section): This is a short and thickened stem, developed from the epicotyl of the seedling. It bears upper leaves and buds.

② Root neck: Developed from the hypocotyl of the seedling, this part lacks leaves and generally does not produce lateral roots.

③ Root (true root): This part is formed by the enlargement of the primary root of the seedling, producing lateral roots. In root vegetables of the Brassicaceae and Chenopodiaceae families, lateral roots are arranged in two rows, while in the Umbelliferae family, they are arranged in four rows. The root head, root neck, and true root together form a unified structure and act as organs for nutrient storage.

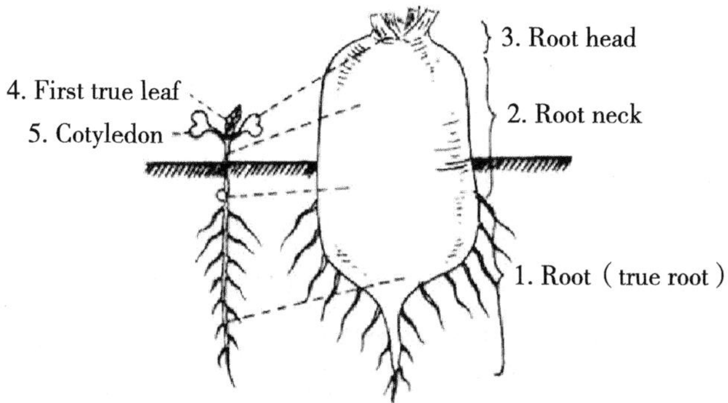

Fig. 5-2 Fleshy Roots of Radishes

(2) Stem: During the vegetative growth period, the stem shortens. In the reproductive growth period, it elongates to form a flowering stem, which can also branch. After entering the growth period, the apical bud produces the main stem, and axillary buds on the main stem can develop into lateral branches. Flowers are borne on the main and lateral branches.

(3) Leaves: Radish plants have two cotyledons, which are kidney-shaped. The first pair of true leaves are spoon-shaped. Rosette leaves grow on the shortened stem. There are two types of leaves: basal leaves and cauline leaves. The leaf clusters grow on the shortened stem during the vegetative growth period, and their shapes, sizes, colors, and spreading patterns vary according to the varieties. The arrangement includes upright, semi-upright, flat, and prostrate types. Upright leaves are suitable for dense planting, while flat leaves are suitable for sparse planting.

(4) Flowers: The flowers are arranged in compound racemes, each flower being complete. The colors range from white and pink to light purple. White-flowered varieties yield white radishes, purple-flowered varieties yield purple radishes, and light purple or white flowers result in red radishes. Radishes are cross-pollinated naturally through insect pollination.

(5) Fruits and seeds: The fruit is a long siliqua that does not easily split upon ripening. Each fruit contains 3–8 seeds. The seeds are irregularly spherical, ranging from light yellow to dark brown, with a weight of 7–15 g per thousand

seeds. The viability of the seeds lasts for 1–2 years, but germination capacity can be maintained for up to 5 years. It is advisable to use fresh seeds for 1–2 years for cultivation.

2. Requirements for environmental conditions

(1) Temperature: Radishes are semi-cold-tolerant vegetables, preferring to grow in mild and cool climates. The optimal temperature for seed germination is 20–25°C. Radish seedlings have a broad temperature tolerance, being able to grow in temperatures as high as 25°C and as low as -2°C, but the growth of mature plants is less adaptable to extreme temperatures. Stem and leaf growth occur in temperatures ranging from 5–25°C, with an optimal range of 15–20°C. For the growth of the fleshy root, temperatures ranging from 6–20°C are suitable, with an optimal range of 18–20°C. The temperature range for vernalization is 1.0–24.6°C, with the easiest vernalization occurring at 1–5°C. Therefore, when planting radishes in spring, it is crucial to select varieties with strong winter hardiness and rapid root formation; otherwise, premature bolting may occur. As for soil temperature, the most suitable range is 20–23°C, with an upper limit of 33°C. The young roots start elongating at 6°C, root hairs develop at 8°C, and rapid elongation occurs at 15–20°C, with 25–28°C as the most favorable temperature range. Increasing soil temperature within the appropriate range can promote the development of the root system.

(2) Light: Radishes require moderate light intensity. Insufficient light can result in slow enlargement of the fleshy root, leading to low yields and poor quality. Radishes are classified as long-day plants. Plants undergoing vernalization under long daylight hours and relatively high temperatures experience rapid flower bud differentiation, and flowering shoot emergence. Therefore, during the spring season when radishes are exposed to long daylight hours, there is a phenomenon called premature bolting.

(3) Water moisture: Radishes prefer moist soil but are intolerant to waterlogging and drought. The ideal soil humidity content for the growth of radish roots ranges from 60%–80%. The air humidity should be maintained at 80%–90%. Radishes have large leaves, resulting in high transpiration rates and significant water

consumption. Additionally, the fleshy roots have high water content and most of the absorbing roots are distributed in the shallow soil layer, making them less efficient in water uptake. Therefore, it's crucial to provide an ample water supply during radish cultivation. Studies have shown that a radish plant weighing 1 kg requires approximately 5 kg of water throughout its life cycle. However, excessive water, especially prolonged waterlogging in the field, should be avoided. Maintaining appropriate soil humidity is essential for radish growth. Excessive humidity can lead to root rot or blackening, while overly dry conditions can intensify the radish's spiciness. Uneven water supply may cause the radishes to crack.

(4) Nutrient conditions of soil: Radish has strict requirements on soil conditions. The planting site for radishes should be rich in organic matter, with good water retention and drainage. Deep-rooted varieties especially demand suitable soil conditions. However, soil that is excessively loose or sandy with poor water retention capabilities can result in radishes becoming hollow and having poor skin quality. In addition, overly dry soil can lead to toughening of the radish flesh and an increase in bitterness. The ideal soil pH value for radish cultivation ranges from 5.3–7.0. Radishes have high nutrient demand during their growth period, requiring a balanced supply of nitrogen, phosphorus, and potassium. Potassium is absorbed in the largest quantities, followed by nitrogen, phosphorus, calcium, and magnesium. During the radish root enlargement period, the absorption of phosphorus and potassium increases rapidly. Therefore, excessive nitrogen fertilization should be avoided during the peak growth stage. The absorption ratio of N, P, and K should ideally be around 2.1 ∶ 1 ∶ 2.5. Adequate boron levels are beneficial for improving radish quality. A deficiency in boron can lead to the development of black root disease.

II. Types and Varieties

In China, there are many varieties of radishes, which can be classified based on their root shapes into long, round, flat round, spindle-shaped, and conical shapes. Depending on the colors of the skin, radishes can be red, green, white, purple, and so on. They are also categorized according to their uses, such as for cooking, processing, and pickling, as well as for eating fresh. Furthermore, radishes are divided into early-, medium-, and late-maturing varieties based on their growth periods. Additionally, they can be classified according to the cultivation seasons, including autumn-winter radishes, winter-spring radishes, spring-summer radishes, summer-autumn radishes, and all-season radishes.

(1) Autumn-winter radishes: They are typically sown in late summer to early autumn and harvested from late autumn to early winter, with a growth period of 60–120 days. These radishes have a wide range of varieties, dominated by medium- to large-sized varieties, and due to favorable growing conditions, they yield high production with excellent quality. They are also resistant to storage and transportation, making them widely cultivated in China. There are three main categories within this type: green radishes, red radishes, and white radishes. Representative varieties include Beijing Xinlimei, Deri No.2, Dahongpao, Tongyuanhong No.1 and No.2, Jinghong No.1, Zhedachang Radish, Meinongzaosheng, Zheluo No.1, etc. In the Yangtze River basin, these radishes are generally sown from the beginning of autumn until the end of the hot summer days, and they are abundantly available in the market from November to December. Varieties cultivated in the southern regions include Huochetou Radish, Wuqing No.1, Zhedachang Radish, Fengguang Radish, Jiuguan Radish, etc.

(2) Winter-spring radishes: They are sown in late autumn to early winter, around October. They overwinter outdoors and are typically harvested from February to April, with a growth period of 120–150 days. These radishes are characterized by their strong cold resistance, strict vernalization requirements, delayed bolting, and resistance to hollowing. The main varieties include Winter-Spring No.1, Winter-Spring No.2, Wuhan Chunbulao Radish, Chengdu Chunbulao Radish, and Ningbai No.3.

(3) Spring-summer radishes: They are typically sown in March to April and harvested in April to May, with a growth period of 40–70 days. These radishes are usually medium-sized, have relatively low yields, and have a short supply period. They also tend to bolt early. To address these challenges, breeding efforts focus on developing high-yielding, cold-tolerant varieties with extended supply periods and

delayed bolting. The main varieties include Spring Red No.1, Spring Red No.2, March Radish, Zuixiantao, Paolihong, Nannong Siji No.1 and No.2, and Spring White Radish.

(4) Summer-autumn radishes: They are sown in early July and harvested in late August to early September, with a growth period of 50–70 days. These radishes are heat-tolerant, resistant to moisture, strong against diseases and pests, less likely to become pithy, and have a relatively short growth period. The main varieties includ Xiakang 40d, Banjiehong, Xinlimei, Double Red No.1, Short Leaf No.13, Dongfang Huimei, and Mid-autumn Red Radish.

(5) All-season radishes: They are mostly small-sized varieties that can be sown throughout the year, with spring sowing being the most common. These radishes mature very early, with a growth period ranging from 20–45 days. They have small and tender roots, making them suitable for raw consumption and pickling. All-season radishes are highly adaptable, being tolerant to both heat and cold. They are less likely to bolt, ensuring good quality, although their yield is relatively low. The main varieties include Shanghai Small Red Radish, Nanjing Yanghua Radish, Jiangjin Yanzhi Radish, Harbin Suanpanzi Radish (also known as Cherry Radish), and Water Chestnut-shaped Radish.

Recently, to meet the demands of the international market, China has successively introduced a variety of high-quality radish cultivars. These include strains such as Baiguang, Shinong R706, Dayizaoguang, Dayichunhetian, Hanyu, Hanbaiyu, Texinbaiyu, Baiyuchun, Meibaichun, Changchun Daikon, and Greenhouse Daikon from Korea. Additionally, from Japan, cultivars such as Yu Daikon, Yuejin Series, Zhentairen, Disease-resistant Rentai, Ideal Series, Tianchun Daikon, April White, and Lichun Daikon have been introduced.

III. Cultivation Season and Methods

In the southern regions of China, radishes can be cultivated throughout the year. In most northern areas, radishes are primarily grown in spring, summer, and autumn, with an emphasis on the cultivation of autumn-winter radishes. Radishes are best rotated with non-cruciferous vegetables. Suitable preceding crops include melons, onions, garlic, eggplants, and legume vegetables. Alternatively, for a grain-vegetable rotation, preceding crops such as wheat, soybeans, and corn can be selected.

IV. Cultivation Techniques

(1) Soil requirements and crop rotation planning: It is essential to choose deep and sandy loam soil and avoid consecutive cultivation of radishes in the same soil. It is recommended to consider planting early-maturing radishes in early autumn after cultivating cucumbers, and late-maturing autumn-winter radishes after planting cowpea.

(2) Land preparation, fertilization, and bedding: It is recommended to avoid planting radishes after cruciferous vegetables and nightshade vegetables such as tomatoes and peppers. It is preferable to choose melons or leguminous vegetables as the previous crops. Land preparation should be meticulous, involving early and thorough plowing. Deep plowing combined with the application of basal fertilizer is crucial. It is crucial to apply 2,500–5,000 kg of well-rotted stable manure per mu, mix the soil and manure thoroughly, and then flatten the surface. In northern regions, it is recommended to create ridges with a height of 20–30 cm and a spacing of 40–60 cm. Alternatively, flat beds with a width of 1.2–1.5 m can be prepared. In southern regions, deep furrows with high ridges are recommended.

(3) Sowing: The ideal sowing period is from early August to early September. Before sowing, it is necessary to carefully select large, plump seeds. These seeds can be sown in holes or in rows. If the soil is dry, it's advisable to water it first and then sow the seeds after the water has permeated the soil. In hole sowing, it is required to dig holes according to the desired plant spacing and sow 5–8 seeds in each hole, ensuring some distance between the seeds. After sowing, it is essential to cover the seeds with soil to a depth of 2–3 cm and lightly press it down. For row sowing, it is crucial to create shallow furrows in the middle of the ridge, sow the seeds evenly in the furrows, cover them with soil, and press down the soil gently. For large-sized varieties, the row spacing should be 40–50 cm, and the plant spacing should be 30–40 cm. For medium-sized varieties, the row spacing should be 20–25 cm, and the plant spacing

should be 15–20 cm.

(4) Field management.

① Thinning seedlings in time, combined with soil cultivation. Thinning should be done based on the principles of early thinning, multiple thinning, and late thinning. The first thinning should be carried out after full germination, leaving a distance of 3–4 cm between seedlings. The second thinning should be done when the seedlings have developed 5–6 true leaves, leaving a distance of 10–12 cm between seedlings. For large-sized varieties, the spacing should be 20–30 cm, and for medium-sized varieties, the spacing should be 15–25 cm. It is advisable to water the seedlings carefully, ensuring the soil remains moist. After rainfall, it's necessary to ensure proper drainage to prevent waterlogging and promote the growth of young seedlings.

② Intertillage and weed control. It is essential to perform intertillage 2–3 times from the seedling stage to the ridge closure stage, ensuring the soil remains loose and well-aerated, preventing compaction. After thinning, it is crucial to combine intertillage with soil cultivation to promote the full development of the fleshy roots. Be careful not to damage the root system during intertillage to prevent forked roots.

③ Topdressing and irrigation. Topdressing should be applied early, and the soil should be kept moist. The principle of radish fertilization is to prioritize basal fertilization, with supplemental topdressing. Basal fertilization promotes root growth, topdressing encourages leaf development, and basal fertilization promotes head formation. It is recommended to water 2–3 times in the early stage to stimulate the growth of rosette leaves, control watering appropriately during the medium and late stages, and perform frequent cultivation to loosen the soil to prevent excessive leaf elongation. After the fleshy roots are exposed, it is crucial to perform a significant topdressing combined with intertillage and irrigation. About 100 kg of wood ash and 30–40 kg of microbial compound fertilizer should be applied between the rows per mu. During this period, it is required to ensure uniform water and fertilizer supply to prevent root cracking. Generally, the plants should be watered every 5–7 days and compound fertilizer should be applied every 2 weeks.

(5) Pest and disease control: Common pests include aphids, cabbage worms, diamondback moths, and cutworms. Radish plants are susceptible to diseases such as black rot, viral diseases, blackheart disease, and downy mildew.

(6) Harvesting: Radishes should be harvested when their fleshy roots are fully swollen, and the leaves start to turn yellowish-green. It's best to harvest on sunny days at the right time. If harvested too early, the yield might be low; if harvested too late, they can become hollow, affecting their quality. When harvesting, it is required to remove the top leaves immediately to reduce moisture evaporation and prevent pithiness. It's advisable to handle radishes gently to avoid bruising, as damaged radishes do not store well.

V. Common Problems and Prevention Measures in Radish Cultivation

1. Premature bolting

Premature bolting refers to the presentation of flower stalk and even premature flowering of radishes before the roots have fully developed. This results in radishes with loose, hollow, and inedible flesh. The main causes include sowing too early, improper variety selection, and inadequate management.

Prevention and control measures: Choose radish varieties with strong cold tolerance, use fresh seeds, sow at the appropriate time, and ensure proper watering, fertilization, and pest and disease control.

2. Deformed roots

Deformed roots, characterized by bending or forking, usually occur in hard or compacted soil where the taproot faces obstacles hindering its growth. Damage to the growing point of the taproot or obstruction in its growth can lead to swollen lateral roots, forming forked fleshy roots. Specific reasons for this phenomenon include shallow soil tillage, hardened soil, or gravel hindering the growth of fleshy roots; using old seeds; damage to the taproot caused during seedling transplantation; application of uncomposted organic fertilizer or excessively concentrated fertilizers; damage to the radicle by pests and diseases.

Prevention and control measures: Select high-quality new seeds, choose deep and loose sandy loam or loamy soil; ensure thorough composting of organic fertilizer; maintain moderate seed sowing density; use direct sowing or transplanting with soil clumps; apply topdressing in appropriate and uniform quantities; irrigate at the right time, and prevent and control pests and diseases in a timely manner.

3. Cracked roots

Cracked roots and fleshy root cracking occur mainly due to an uneven supply of nutrients and water during the period of fleshy root expansion. This can happen if there is dryness in the early stages followed by excessive watering later, or vice versa—excessive watering in the early stages followed by dryness later. Additionally, excessive nitrogen fertilizer, improper management practices, and delayed harvesting can also lead to root cracking.

Prevention and control measures: Timely irrigation during dry periods, proper drainage when there is excess moisture, appropriate supplementation of nitrogen fertilizer at the right time and in the right amount, enhanced management practices, and timely harvesting.

4. Hollow heart

Hollow heart, also known as pithiness, refers to the phenomenon where the central part of the xylem in fleshy roots becomes hollow. This mainly occurs during the period when the roots are exposed. During the peak growth stage of fleshy roots, when plant respiration and transpiration are vigorous, the roots expand rapidly, consuming a large amount of water and nutrients. If there is high temperature and drought, insufficient watering, some parenchyma cells in the xylem, which are far from the vascular tissues, lack nutrients and water supply. This leads to the loss of sugars in the cells, reduced soluble solid content, and the formation of intercellular space, resulting in a hollow heart. Fast-growing varieties with soft flesh can be affected by several factors, including early sowing leading to excessive vegetative growth, early bolting and flowering, over-application of nitrogen fertilizer, insufficient potassium fertilizer, delayed harvesting, overly dry covering soil during storage, high-temperature dry conditions, and prolonged storage period. All these factors can cause radishes to lose a significant amount of water and result in hollow heart.

Prevention and control measures: Select varieties with dense and solid flesh; ensure timely drainage; apply fertilizer evenly; supply sufficient water and nutrients; harvest in a timely manner; supplement with potassium fertilizer; implement proper dense planting, and sow at the appropriate time.

5. Blackheart

Blackheart is a condition where the extraradical appearance of the fleshy root appears normal, but the interior turns black due to difficulties in respiration caused by lack of oxygen. The primary cause is related to soil conditions, such as soil compaction, excessive soil moisture, and insufficient composting of organic fertilizer.

Prevention and control measures: Strict control of soil compaction, effective water management to prevent excessive soil moisture, application of well-composted organic fertilizer, and timely prevention and control of diseases like black rot.

6. Spiciness and bitterness

Radishes contain mustard oil in fleshy roots. When the mustard oil content is moderate, radishes taste good. However, if the content is too high, the spiciness intensifies. Hot and dry weather, as well as insufficient application of organic fertilizer, can increase the mustard oil content in radishes. Additionally, in hot weather or when excessive nitrogen fertilizer is applied without sufficient phosphorus and potassium fertilizer, fleshy roots can produce a nitrogen-containing alkaloid compound called momordicin, leading to a bitter taste.

Prevention and control measures: Select high-quality varieties; pay attention to the application of organic fertilizer; supplement with phosphorus and potassium fertilizer; sow at the appropriate time; manage temperature and humidity, and prevent and control pests and diseases in a timely manner.

「**Summary of Sub-context**」

This sub-context primarily focuses on the cultivation techniques of radishes, with a specific emphasis on land preparation and field management.

「Expanded Knowledge」

Taboos for Eating Radishes

Radishes should not be consumed together with oranges. Radishes contain a large amount of carotene, which is beneficial for eyesight and eye health. However, when eating together with oranges, it may cause problems. As radishes produce a substance called thiocyanate, which is anti-thyroid. If consumed in large quantities with fruits like oranges, apples, grapes, etc., the flavonoids in these fruits, after being broken down by bacteria in the intestines, will convert into thiocyanate, which inhibits thyroid function and can lead to thyroid enlargement.

Symptoms of the disease include: alternating pale and yellow complexion, fluctuating body temperature, vomiting, cold sweating, weakness, etc.

Sub-context 2 Production of Carrot

Carrot, also known as red carrot, yellow carrot, and clove carrot, is a biennial herbaceous plant belonging to the *Daucus* genus of the Apiaceae family. It is native to Afghanistan. Carrots have extremely high nutritional value and are often called the "little ginseng". They have bright colors, a crisp texture, and a delicious taste. The fleshy roots of carrots are rich in nutrients such as carotene, sugar, starch, potassium, calcium, phosphorus, iron, and other minerals. Carrots, in particular, have a high content of carotene, which is 5–7 times higher than that of tomatoes. Carotene is converted into vitamin A in the human body through digestion and hydrolysis, and it can prevent night blindness and respiratory diseases. Carrots can be consumed fresh or processed into products such as carrot juice, carrot sauce, dehydrated carrots, and refined carotene. Carrots are highly adaptable, easy to cultivate, have few pests and diseases, and can be stored and transported for a long time. They play a significant role in regulating market supply. There has been a sharp increase in international demand for carrots, and currently, China has the world's largest carrot cultivation.

I. Biological Characteristics

1. Botanical features

(1) Roots: Carrots have a tap root system, characterized by deep roots that can reach depths of over 2 m. They spread horizontally between 1 m and 1.5 m, with the main root system located within the soil layer of 20–90 cm. Carrots exhibit a good resistance to drought. The fleshy root serves as the edible organ.

(2) Stems: During the vegetative growth period, the stems of carrots are short and compact, bearing leaves. In the reproductive growth phase, they develop flowering stems, which can branch extensively. The plant has a strong branching ability. Lateral branches can emerge from every node of the main stem, and secondary branches can develop from these lateral branches.

(3) Leaves: Leaves are clustered on the short and compact stems and are tripinnate. The petioles are slender, and the leaves are rich green in color. They have a small surface area densely covered with fine hairs, and exhibit drought resistance.

(4) Flowers: Carrots have compound umbel, with each inflorescence bearing thousands of florets and complete flowers. The flowers are white or pale yellow and exhibit monoecious characteristics. Cross-pollination occurs through insect-mediated pollination, making them susceptible to natural hybridization.

(5) Fruits and seeds: Carrots produce mericarps that are yellowish-brown in color. The fruits are used as planting material. The surface of the fruit has longitudinal furrows and splits into two parts when ripe. The pericarp is leathery, covered with prickly hairs, and contains volatile oils. It has poor water absorption capabilities and weak emergence ability. The germination rate is low, typically around 70%. Carrot seeds lack endosperm and weigh from 1.1–1.5 g per thousand seeds. The seed's viability lasts for 2–3 years.

2. Requirements for environmental conditions

(1) Temperature: Carrots are semi-cold-tolerant vegetables with slightly better

cold and heat resistance compared to radishes. Therefore, they can be sown earlier and harvested later than radishes. The optimum germination temperature for seeds is 20–25°C. Carrot seedlings are tolerant to both low and high temperatures, and they usually emerge within about 5 days under suitable temperatures after sowing. The ideal temperature for leaf growth is 23–25°C during the day and 15–18°C at night. For the enlargement of fleshy roots, the suitable daytime temperature is 18–23°C, nighttime temperature is 13–18°C, and soil temperature is 16–18°C. When the temperature exceeds 28°C or drops below 3°C, the enlargement of fleshy roots stops. The color of fleshy roots is sensitive to low temperatures; the temperature range of 15–21°C is favorable for carotene formation, resulting in well-colored roots.

Carrots belong to the green-type biennial plants that require vernalization. After approximately 10 leaves have developed, the plant undergoes vernalization under low-temperature conditions of 2–6°C for 50–100 days. In some southern regions, a few varieties can undergo vernalization under conditions of seed germination and relatively higher temperatures.

(2) Light: Carrots thrive in abundant sunlight. With sufficient light, photosynthesis is strong, assimilation products are abundant, and fleshy roots expand rapidly. Inadequate sunlight leads to poor plant growth, hindered enlargement of fleshy roots, low yields, and inferior quality. Carrots are considered long-day plants, and plants that have undergone vernalization require daylight conditions of over 14 hours to progress through the flowering and fruiting stages.

(3) Water moisture: Carrots have well-developed root systems and are one of the most drought-tolerant vegetables among root crops. During the germination period, soil moisture is essential to facilitate seedling emergence. In the stage of fleshy root expansion, carrots require ample water. However, water supply should be balanced to prevent over-drying or excessive moisture, which can lead to root cracking or the formation of low-quality roots. During the seedling and rosette stages, a combination of promotion and control of water is required.

(4) Soil nutrition: Carrots have similar soil requirements to radishes. They thrive in deep, loose, well-drained sandy loam soil or loam with a pH range of 6–8.

In the early growth stage, carrots require a higher nitrogen content in the soil. They also have a certain demand for phosphorus. As the fleshy roots expand, the need for potassium increases. The ratio of N, P, and K is 2.5 : 1 : 4. Foliar spraying with boron fertilizer can enhance carrot quality and contribute to increased yields.

II. Types and Varieties

Carrots can be classified into three types based on the shapes of their fleshy roots: long conical, short conical, and long cylindrical. Additionally, carrots can be categorized into three classes according to the colors of their fleshy root skin: red, yellow, and purple. When selecting varieties, it is advisable to choose excellent cultivars with small leaf clusters, large and well-shaped fleshy roots, smooth surfaces, small cores, fine and dense flesh, juiciness, high sugar, and carotene content, resistance to cracking or forking, resistance to bolting, high yields, and disease resistance.

(1) Long conical type: Mostly medium- to late-maturing varieties, these carrots are red or purplish-red in color. They have slender, elongated roots with pointed tips, offering a sweet taste and good storage capabilities. The fleshy roots are typically 25–50 cm in length. Common varieties include Shantou Red, Chengdu Xiaoyingzi, and Dali Red Carrot.

(2) Short conical type: Early-maturing and heat-tolerant, these carrots have lower yields. They tend to bolt later when cultivated in spring. The fleshy roots are about 15–20 cm long. Well-known varieties in this category include Shandong Yantai Five Inches, Jiangsu All-season Carrot, and Shanxi Erjin Red Carrot.

(3) Long cylindrical type: Late-maturing, these varieties have long, slender roots with thick shoulders and slightly broader tips. The fleshy roots are typically 30–40 cm in length. Popular varieties include Shanghai Long and Red Carrot, Zhejiang Dongyang Yellow Carrot, Heitian Five Inches, and Hangzhou Red Heart.

III. Cultivation Seasons and Crop Rotation

Carrots are typically cultivated in two seasons: spring and autumn, with the autumn season being the primary cultivation period. In some regions, carrots are grown in three seasons: spring, summer, and autumn. Additionally, there are methods like spring cultivation under thin film covering. Carrot seedlings have slow growth, so the sowing time in autumn should be earlier than that of radishes. In the northeastern and high-altitude regions, planting usually starts in June. In the northwestern and northern regions, sowing occurs in mid- to late July. In the middle and lower reaches of the Yangtze River, planting is generally done from late July to early August. In southern China, sowing typically takes place from late September to late October.

For autumn carrots, suitable preceding crops include cabbage, tomatoes, cucumbers, garlic, onions, common beans, cowpeas, wheat, and peas. Spring carrots are usually rotated after crops like spinach, loose-headed Chinese cabbage, and autumn cabbage.

IV. Cultivation Techniques

In recent years, due to the increasing demand for off-season vegetables and higher market prices, along with the need to meet export requirements, many regions have adopted off-season cultivation. This practice has yielded favorable economic benefits.

1. Soil preparation and fertilization.

Carrots have strong drought tolerance and are susceptible to waterlogging. Therefore, it's essential to choose deep, loose, fertile, and well-drained loam or sandy loam soil. Prepare the soil meticulously by combining fine tillage and deep plowing. 3,000–4,000 kg of well-rotted stable manure and 30–50 kg of microbial composite fertilizer can be applied per mu. Plow the soil to a depth of 25–30 cm to ensure thorough mixing of organic matter and soil. Level the soil finely and create ridges for cultivation. Maintain ridge spacing at 60 cm with a height of 20 cm. Plant in double rows with a spacing of 11 cm × 20 cm (or 40 cm) between plants, resulting in 20,000 plants per mu.

2. Variety selection

Choose early-maturing, disease-resistant, cold-tolerant, high-yielding varieties that are less likely to bolt during spring cultivation. Some suitable varieties include

Heitian Five Inches, All-season Carrot, Jinghong Five Inches, and Chunzao Red No.2. Among these, Heitian Five Inches is widely cultivated, but it's important to note that it is suitable for autumn planting, so its adaptability should be taken into consideration.

3. Seeding

The fleshy roots of carrots prefer to grow in cool and mild climate conditions. Although carrot seedlings are more heat and drought-resistant than radishes and have a longer growth period, their tolerance is still limited. Therefore, it is possible to take advantage of the heat-resistant characteristics of carrot seedlings to sow spring carrots earlier. In the Yangtze River basin, spring carrots are typically direct-seeded in early spring, around March to April, and harvested for market in May to June. It is advisable to sow when the temperature at a depth of 5 cm in the soil stabilizes at 6–8°C or higher. Pre-soaking the seeds to promote germination and enhance seedling emergence speed is recommended. Broadcasting or row planting methods can be employed, with broadcasting being more commonly used.

4. Field management

During the seedling stage, which coincides with the hot and rainy season, weeds tend to grow rapidly. It is crucial to perform timely intertillage and weed removal. Once the seedlings have emerged uniformly, thinning should be conducted at the right time to remove overcrowded, miscellaneous, weak, and diseased seedlings to prevent overcrowding and ensure adequate sunlight. Thinning should be done when the seedlings have 1–2 true leaves for the first time, 3–4 true leaves for the second time, and 5–6 true leaves for the final thinning. Plant spacing should be around 10 cm for small varieties and 13–15 cm for large varieties. Before emergence, the soil should be kept moist to facilitate seedling emergence. After rain, it is recommended to ensure timely drainage. During the peak leaf growth stage, appropriate water control is necessary to prevent excessive elongation of seedlings. During the period of root enlargement, the soil should be kept moist, maintaining soil humidity at 60%–80%. In the later stages of growth, it is required to stop watering. When irrigating, avoid the hot noon period; it is crucial to preferably water in the early morning or late evening to prevent root cracking.

Carrots have a long growth period. In addition to the initial basal fertilizer, topdressing should be done 2–3 times. The first round of fertilization should be applied 20–25 days after thinning, and the second round should be applied after the final thinning. During the period of root enlargement, an appropriate amount of well-rotted compost or phosphorus and potassium fertilizer can be applied to facilitate the formation of meaty roots.

5. Harvesting

Carrots should be harvested when their roots are fully enlarged and mature. Typically, they can be harvested around 90–100 days after sowing. Signs of maturity include the cessation of leaf growth and no new leaves emerging. Harvesting too late can result in the carrots becoming tough and a decline in market quality.

6. Pest and Disease Control

Main diseases include soft rot, black rot, downy mildew, and leaf blight. Main pests include aphids and celery worms. 20% neem oil, 20% chlorpyrifos, 10% bromophos, and 5% fluoroimide can be employed for pest control.

V. Common Cultivation Issues and Their Prevention Strategies

Common problems in the cultivation of spring carrots include premature bolting, forking, root cracking, bending, lumpy growth strumae, green shoulders, long root hairs, color variations, and thickening of the central core of the fleshy root. Solutions to issues such as premature bolting, forking, root cracking, and bending can be referenced from the content relating to radishes.

1. Strumae

When the lateral roots of carrot fleshy roots develop, the surface bulges into lumpy formations, affecting the quality of the produce. The main causes are heavy and poorly-draining soil in the cultivation area, and excessive fertilization, especially an overdose of nitrogen, leading to excessive growth and rapid swelling of the fleshy roots.

Prevention and control measures: Use well-draining sandy loam soil with deep layers, good aeration, and appropriate fertilization. It's crucial to avoid excessive nitrogen fertilizer during the period of fleshy root expansion.

2. Green shoulders

In the later stages of plant growth, insufficient soil covering or shallow soil depth can lead to the exposure of the shoulders of fleshy roots. The main reasons include poor growth conditions, reduced stem and leaf development due to pests and diseases, and high temperatures and dry conditions in the middle and late stages of growth, leading to soil erosion during heavy rainfall.

Prevention and control measures: Select soil with sufficient depth for cultivation, pay attention to pest and disease control in the field management, maintain uniform soil moisture, and enhance intertillage and soil cultivation practices.

3. Long taproots

Long taproots in carrots are mainly caused by compacted soil or poor drainage, resulting in poor aeration.

Prevention and control measures: Choose appropriate soil, provide proper irrigation, deep plowing, fine harrowing, and loosening the soil through intertillage.

4. Color variation

Color variation in carrots occurs primarily due to shallow plow layers during cultivation, lack of attention to soil preparation during the root enlargement period, or late planting, leading to accumulation obstruction of carotene and lycopene in the high-temperature period of July to August. This obstruction results in color variations such as whitening or yellowing.

Prevention and control measures: Deep plowing, fine harrowing, loosen the soil through intertillage, hill planting, and other methods to deepen the carrot's color, smooth the skin, and improve carotene content. Additionally, applying potassium and magnesium can enhance carotene levels and improve the root's color.

5. Thickening of the central core in the root's flesh

The thickening of the central core in the root's flesh primarily occurs due to factors such as the varieties of carrots, excessive plant spacing, and excessive nitrogen fertilizer application.

Prevention and control measures: Choose high-quality carrot varieties, maintain appropriate spacing between plants, and moderate the application of nitrogen fertilizer.

「**Summary of Sub-context**」

This sub-context primarily focuses on the cultivation techniques of carrots, focusing on soil preparation and field management of carrots.

「**Expanded Knowledge**」

The Therapeutic Value of Carrots in Diet

According to research by American scientists, consuming two carrots per day can reduce cholesterol levels in the blood by 10%–20%. Eating three carrots daily is beneficial in preventing heart diseases and tumors. In traditional Chinese medicine, carrots are considered sweet in taste, neutral in nature, and possess various medicinal properties such as tonifying the spleen and stomach, nourishing the liver and improving eyesight, clearing heat and detoxifying, strengthening yang and tonifying the kidneys, promoting rash eruption, and relieving cough and asthma. Carrots can be used to treat conditions such as gastrointestinal discomfort, constipation, night blindness (due to the action of vitamin A), sexual dysfunction, measles, whooping cough, and malnutrition in children. Carrots are rich in vitamins and have a mild and sustained diaphoretic effect. They can stimulate skin metabolism, enhance blood circulation, and contribute to smoothing and glowing skin. Carrots are particularly effective in skincare. They are also suitable for individuals with dry, rough skin or those suffering from conditions like seborrheic dermatitis, blackheads, acne, or keratosis-type eczema.

「**Reviewing and Thinking Questions**」

I. Explanation of Terms

1. Premature bolting.
2. Pithiness.
3. Cracking and shouldering.

II. Gap Filling

1. According to the anatomical structure of fleshy roots, the main edible part of the radish is ________, and the main edible part of the carrot is__________.

2. The external morphology of root vegetables can be divided into three parts: ________, ________, and________, which are developed from________,________, and________, respectively.

3. Excessive spiciness in fleshy roots is due to the high content of________, and bitterness is caused by the presence of________ in the fleshy root.

4. The phenomenon where the central part of the fleshy root becomes hollow is called________, which is related to________, ________, and other factors.

5. In the early growth stage, carrots require a lot of________ fertilizer. They also have a certain demand for________ fertilizer. During the period of fleshy root enlargement, they need more________ fertilizer. Spraying________ fertilizer on the leaves can improve the quality of carrots and has a certain effect on increasing yields.

6. The main cause of carrot fleshy root cracking is________.

7. The color of the fleshy root of carrots is sensitive to________.

8. Carrots are________herbaceous plants belonging to the ________ genus of the ________ family.

III. Choice Questions

The main reason for radish cracking is ( ).

A. Shallow cultivation layer

B. Application of uncomposted organic fertilizer

C. Uneven water supply

D. Bricks in the soil

IV. Thinking and Answering

1. Reasons and prevention measures for the spicy taste, bitterness, pithiness, root cracking, and premature bolting of radishes.

2. Describe the growth and development patterns of radishes and spring radish cultivation techniques.

3. How can year-round cultivation of radishes be achieved?

4. Describe the field management techniques for carrots.

5. What are the fertilizer requirements for root vegetables?

Learning Context 6 Production of Leafy Vegetables

「Learning Objectives in This Context」

Understand the main types, common growth characteristics, and cultivation methods of leafy vegetables, and grasp the biological characteristics, variety types, cultivation seasons, and crop rotation arrangements of major leafy vegetables such as lettuce, celery, and spinach.

「Analysis of Tasks in This Context」

Master tasks such as sowing and seedling raising, land preparation, mulching film, transplanting, fertilizer and water management, pest and disease control, harvesting, and post-harvest treatment for major leafy vegetables such as lettuce, celery, and spinach.

「Introduction」

Leafy vegetables are diverse and indispensable on people's dining tables. They include common varieties such as lettuce, spinach, amaranth, and celery. In the production process, they are often cultivated in small varieties, with cultivation occurring on a small scale and limited area.

Overview

Leafy vegetables primarily consist of tender green leaves, petioles, and young stems, making them fast-growing vegetables consumed for their edible parts. This category includes a wide range of species, genera, and families, exhibiting diverse morphologies, structures, and flavors. Leafy vegetables are highly adaptable, have short growth cycles, and offer flexible harvest periods. They play an irreplaceable role in ensuring a year-round balanced vegetable supply, optimizing variety combinations, increasing crop rotation rates, enhancing per unit area yields, and improving economic efficiency.

China boasts a rich varieties of leafy vegetables, with abundant resources. Worldwide, there are over 15 families and more than 40 species of cultivated leafy vegetables. In China, there are more than 20 species belonging to 13 families (Fig. 6-1). The main varieties include lettuce, celery, spinach, amaranth, cilantro, fennel, leaf mustard, winter radish, *Ixeris denticulata*, shepherd's purse, chicory, sweet wormwood, purslane, *Atriplex hortensis*, water spinach, amaranth, crown daisy, *Basella albal.*, vegetable alfalfa, and *Gynura bicolor*, etc. These vegetables are primarily consumed for their tender leaves, stems, or shoots. Certain varieties, such as celery, are primarily valued for their petioles, while others, like lettuce, are primarily appreciated for their stems.

Leafy vegetables are rich in various vitamins and minerals, particularly abundant in nitrogen compounds, making them highly nutritious. Among these, certain varieties such as shepherd's purse, Chinese mallow, vegetable, alfalfa, purslane, cilantro, amaranth, and spinach contain more than 30 mg of vitamin C per 100 g. Therefore, it is recommended to consume 400–500 g of leafy vegetables daily to meet the body's vitamin C requirements. Leafy vegetables also have a high content of carotenes, including varieties like spinach, celery, Chinese mallow, purslane, cilantro, fennel, shepherd's purse, vegetable alfalfa, etc. These vegetables contain over 2 mg of carotenes per 100 g, making them an excellent source to fulfill the body's carotene needs. Furthermore, leafy vegetables contain other essential nutrients such as folic acid, choline, calcium, iron, and phosphorus. They are particularly crucial for pregnant and lactating mothers. Leafy vegetables are known for their therapeutic and medicinal properties, earning them the nickname "green elves" abroad, underscoring their significant role in promoting human health.

Fig. 6-1 Different Leafy Vegetables

Leafy vegetables have the following common characteristics in biological characteristics and cultivation techniques.

Leafy vegetables can be categorized into two groups based on their temperature requirements. The first group originates from subtropical regions and prefers mild climates. These leafy vegetables thrive in cool temperatures at 15–20°C. They are suitable for autumn planting with autumn harvest or spring planting with spring harvest, or autumn planting for the next year's spring harvest. Examples include lettuce, celery, spinach, amaranth, and cilantro. When cultivated in cool and chilly conditions, they yield high-quality produce. However, under high temperatures or hot and dry conditions, the quality may decrease. For instance, spinach leaves may become smaller and thinner, and the taste might turn bitter. Lettuce leaves may also become smaller and rougher, sometimes acquiring a slightly bitter taste. The second group of leafy vegetables originates from tropical regions and prefers warmth but is sensitive to cold. They grow best in temperature range of 20–25°C and cease growth below 10°C. These varieties are vulnerable to frost and can easily be damaged by freezing temperatures. Examples include amaranth, water spinach, and purslane. However, they are relatively tolerant to high summer temperatures. These vegetables are suitable for spring planting with summer harvest or summer planting with summer harvest. They play a crucial role in increasing the varieties of leafy vegetables available during the summer, particularly in addressing the scarcity of produce in early autumn.

Most leafy vegetables have shallow root systems and are planted densely per unit area. Due to their rapid growth and short growth period, they produce a larger quantity of products per unit weight in a given period. As a result, they absorb a considerable amount of nutrients from the soil and require higher soil and water-fertilizer conditions. Both basal fertilizers and topdressings should be fast-acting, and frequent and light applications are necessary to meet the continuous growth requirements. Sufficient nitrogen fertilizer results in tender and juicy leaves with fewer fibers. In contrast, insufficient nitrogen leads to stunted plants with fewer and yellowish, rough leaves, rendering them inedible.

Sub-context 1 Production of Lettuce

Lettuce, an annual or biennial vegetable crop, belongs to the *Lactuca* genus of the Asteraceae family. It can be classified into leaf lettuce and stem lettuce based on the edible parts. Leaf lettuce varieties include lettuce, mizuna, and oak leaf lettuce. Stem lettuce is also known as asparagus lettuce, stem lettuce, or Romaine lettuce. Lettuce originated along the Mediterranean coast and evolved from wild species. Lettuce is widely cultivated in various regions of China. In the Yangtze River basin, it is one of the main vegetables supplied from March to May. In warm regions, utilizing different varieties for staggered sowing and harvesting allows nearly year-round supply. Leaf lettuce is cultivated worldwide and is mainly found in Europe and the Americas. In China, it was previously commonly grown in Guangdong, Guangxi, and especially Taiwan. Since the 1990s, leaf lettuce has been cultivated in major and medium-sized cities across the country.

I. Biological Characteristics

1. Botanical features

(1) Roots: Lettuce has a tap root system with a taproot that can grow up to 150 cm long. After transplanting, the main root is mostly distributed within the soil layer of 20–30 cm. It develops numerous lateral roots and well-developed fibrous roots.

(2) Stems: Lettuce stems are short and stubby. However, in the case of celtuce, after the formation of rosette leaves, the stem elongates and swells to form a shoot. This shoot grows from the stalk developed from hypocotyl and flower stalk. The outer appearance of the stem can be green, greenish-white, purplish-green, or purple. Internally, the stem is fleshy and can have various colors like green, yellowish-green, or greenish-white.

(3) Leaves: The leaves emerge directly from the base of the plant and are arranged alternately on the short stem. They can be smooth or wrinkled and come in various colors such as green, yellowish-green, or greenish-purple. The shapes of

the leaves include lanceolate, elongated oval, and elongated inverted ovate. In leaf lettuce, after the formation of rosette leaves, the central leaves, depending on the varieties, can form spherical, flat spherical, conical, or cylindrical leaf heads. The leaf margins can be wavy, shallowly lobed, or serrated.

Lettuce contains milky white latex, composed of sugars, mannitol, resin, proteins, lactucin, rubber, and various mineral salts.

(4) Flowers: Lettuce flowers are arranged in cone-shaped capitulum, with approximately 20 florets in each inflorescence. They are pale yellow in color. Lettuce is primarily self-pollinating, but occasionally, it can also be cross-pollinated by insects. Generally, seed maturity occurs 11–15 days after flowering.

(5) Fruits and seeds: The fruit of lettuce is classified as an achene in botanical terms. It is small and elongated, ranging in color from grayish-black to yellowish-brown. When mature, it has a parachute-like pappus at the tip, facilitating wind dispersal. Seed collection should ideally be done before the seeds are dispersed by the wind to prevent losses. Its weight of one thousand seeds is 0.8–1.2 g.

2. Requirements for environmental conditions

(1) Temperature: Lettuce prefers cool temperatures and is intolerant to high heat, but it can tolerate light frost injury. The minimum temperature required for seed germination is 4°C, but germination occurs more slowly at this temperature. The optimal germination temperature range is 15–20°C, and seeds can germinate within 4–6 days at these temperatures. If the temperature exceeds 30°C, the seeds enter a dormant state, hindering germination. Therefore, during hot seasons, seeds need to be subjected to low-temperature treatment. For instance, soaking the seeds for germination at temperatures between 5°C and 18°C results in robust germination.

Lettuce has temperature requirements in different growth stages. Lettuce seedlings can tolerate temperatures as low as -5– -6°C. However, the cold tolerance diminishes as the plant matures. The optimal temperature range for seedling growth is 12–20°C. Even when the daily average temperature reaches around 24°C, the seedlings continue to grow vigorously. During the stem and leaf growth stage, the suitable temperature range is 11–18°C. Lower night temperatures and significant day-night temperature variations are beneficial for the thickening of the stems. If the daily average temperature exceeds 24°C and the night temperature remains above 19°C for an extended period, the plants are prone to elongation, resulting in slender stems. Mature lettuce plants are sensitive to temperatures below 0°C and may suffer frost damage or die at such low temperatures. Lettuce plants require higher temperatures during the flowering and grain-filling stages. In the temperature range of 22.3–28.8°C, higher temperatures lead to shorter time intervals between flowering and seed maturation. Temperatures below 15°C adversely affect flowering and seed-setting processes.

Heading lettuce has a narrower temperature tolerance range than celtuce; It is neither cold-resistant nor heat-resistant. The suitable temperature range for heading lettuce during the heading stage is 20–22°C during the day and 12–15°C at night. If the average daily temperature exceeds 20°C, it can lead to poor growth, elongation, difficulty in forming leaf heads, or even necrosis and rotting of the inner leaves due to excessively high temperatures inside the head. Non-heading lettuce has an adaptability to the temperature that falls between that of celtuce and heading lettuce.

(2) Light: Lettuce is a photophilic plant that thrives in abundant sunlight. With ample light, it grows robustly, with thick leaves and thick tender stems. Lettuce is a long-day plant in terms of development. Its seeds require light for germination, and proper scattered light can promote sprouting.

(3) Water moisture: At each stage of growth, lettuce requires appropriate moisture to grow normally. During the seedling stage, do not excessively dry or wet. During the bulb formation stage, water should be controlled timely, and thinning should be carried out to promote the deep growth of the root system, allowing the rosette leaves to develop fully. During the period when the stem of celtuce enlarges or during the heading stage, it is essential to provide sufficient water and nutrients to promote the full development of the edible parts of the plant.

(4) Soil and nutrition: Lettuce has a high demand for oxygen in its root system. It develops rapidly in soil rich in organic matter, with good water retention and fertility, such as loam or sandy loam soil. This type of soil is conducive to the

absorption of water and nutrients by the plant. Heading lettuce prefers slightly acidic soil, with a pH value of around 6, which is considered the most suitable for its growth. Lettuce has high soil nutrient requirements, with a particular emphasis on nitrogen, which is crucial at all stages of its growth and should never be lacking. Lettuce has high soil nutrient requirements, with a particular emphasis on nitrogen, which is crucial at all stages of its growth and should never be lacking. During the heading and stem swelling stage of celtuce, it is essential to maintain a balance between nitrogen and potassium nutrition while also supplying phosphorus.

II. Types and Varieties

Lettuce can be divided into two types based on the edible parts: Stem lettuce and leaf lettuce.

1. Stem lettuce

Stem lettuce, namely celtuce, is a variant of the *Lactuca* genus capable of forming fleshy stems. It can be further categorized into two types based on the shapes of the leaves: pointed leaf and round leaf stem lettuce. Within these types, there are distinctions in stem color, including white (greenish-white outer skin), stem lettuce (pale green outer skin), and purple skin shoots (purplish-green color).

Pointed leaf stem lettuce: The leaves are lanceolate, tapering at the tip, forming small clusters. The internodes are relatively sparse, and the leaf surface is smooth or slightly wrinkled, appearing in green or purple. The fleshy stem is cylindrical, thicker at the base and thinner at the top. This variety matures later and is heat-tolerant during the seedling stage, suitable for autumn or winter cultivation. Their main varieties include Sichuan Zhengxing No. 3 Stem Lettuce, Sichuan Zhongdu Brand Dongqing Stem Lettuce, Dong Stem Lettuce, Bailu Stem Lettuce, Hangzhou Pointed Stem, Shanghai Pointed Stem, Nanjing White Fragrance, Hunan Large Pointed Stem, Chengdu Pointed Stem, Chongqing Wannian Stake, etc.

Round leaf stem lettuce: The leaves are long ovoid, slightly round at the top, with more wrinkles on the leaf surface. The leaf clusters are larger, and the internodes are dense. The stem is thick (thicker in the middle and lower parts, gradually tapering towards the ends). This variety matures early, and has strong cold tolerance, but is not heat-resistant. It is mainly cultivated as overwintering or spring stem celtuce. Their main varieties include Sichuan Zhongdu Brand Autumn Stem Lettuce, Hangzhou Round Leaf, Shanghai Small Round Leaf and Large Round Leaf, Nanjing Purple Fragrance, Hubei Xiaoxian Stem Celtuce, Hunan Luochui Stem Lettuce, Chengdu Er Bai Pi, Guasi Red, etc.

2. Leaf lettuce

Leaf lettuce includes three categories: heading lettuce, curled lettuce, and loose-leaf lettuce.

Also known as long-leaf lettuce, loose-leaf lettuce has entire or serrated leaves. The outer leaves are upright and generally do not form heads or have loose cylindrical or conical heads. This type is widely cultivated in Europe and America. It can be categorized into green and purple varieties based on leaf colors. The former include Nongyou Cuihua, Guangdong Glass Lettuce, and Dengfeng Lettuce, while the latter include Honghuohua and Taiyanghong.

Curled lettuce has deeply lobed leaves with wrinkled surfaces, and it may form loose heads or remain non-heading.

During the seedling stage, do not excessively dry or wet. Leaves of heading lettuce are entire, serrated, or deeply split, with smooth or wrinkled leaf surfaces. The outer leaves are spread out, and the heart leaves form leaf heads. The head can be round, oval, or conical in shape. Key varieties include Big Lake, Green Lake, Big Boston, and Caesar cultivated in the United States.

III. Cultivation Season and Methods

Lettuce prefers cool and temperate climates, and the optimal temperature range for stem and leaf growth is 11–18°C. Mature plants are not frost-resistant, and under long daylight and high-temperature conditions, they tend to bolt. In colder regions, spring cultivation is the primary practice during winter. In relatively warmer areas, apart from spring and autumn cultivation, it is possible to extend cultivation by either starting earlier in spring or delaying planting in autumn. Currently, lettuce cultivation is divided into four seasons: spring, summer, autumn, and winter. In recent years, with the utilization of techniques such as winter protection for cold resistance and shade and rain cover for temperature reduction in summer, lettuce cultivation can be staggered, ensuring year-round supply. This not only enriches the varieties of products in the market but also increases economic benefits.

Leaf lettuce has strong adaptability and can refer to the cultivation season of stem lettuce. Heading lettuce has a narrower temperature tolerance range and is not resistant to extreme low or high temperatures. In regions south of the Yangtze River, it is sown in autumn or winter and harvested in spring, or sown in autumn and harvested in winter. In Guangzhou, where winters are mild with minimal frost damage, planting can occur from August to February of the following year, with harvesting taking place from September to April of the following year. However, the main cultivation season is from October to December for sowing, with harvesting from December to March of the following year. In the southwestern mountainous areas, it can be planted in spring for summer harvest or in autumn for winter harvest. In recent years, in some areas of the Yangtze River basin, experiments have been conducted. Based on the variations in temperature during different periods of the year and the differences in varieties, protective measures such as shading to prevent rain and reducing temperature and humidity during the summer (from late June to early September) and protective measures such as multiple layers of covering during winter (from November to early April of the following year) have been taken. By using small batches of multi-stage sowing (approximately 20 planting periods throughout the year, with planting every 15–20 days), year-round production and balanced supply can be achieved.

IV. Cultivation Techniques

1. Cultivation of stem lettuce

Depending on the sowing periods, stem lettuce can be divided into spring stem lettuce and autumn stem lettuce.

(1) Sowing and seedling cultivation. For cultivating stem lettuce, it is common to start with seedling cultivation before transplantation. To cultivate robust seedlings, it is essential to select high-quality seeds. Uniform germination, strong vitality of seedlings, and high seedling survival rate can lead to higher yields and

save seed usage. Additionally, proper spacing during sowing is crucial to prevent overcrowding, which can result in elongated hypocotyls and tender tissues. Generally, the seeding rate is around 11.25 kg per hectare of seedbed, which can be used to transplant 39–40 hectares of field. The seedbed should be prepared with well-rotted compost and manure, supplemented with appropriate phosphorus and potassium fertilizers. Thinning of seedlings should be done 1–2 times to ensure healthy growth when overcrowding occurs. Transplanting is typically done when the seedlings have 4–5 true leaves, with autumn stem lettuce being slightly larger than spring stem lettuce at transplanting.

(2) Soil preparation and field planting. Stem lettuce has shallow root systems, so it is essential to choose fertile soil with good water retention and fertility for cultivation. Stem lettuce is relatively adaptable to soil acidity and alkalinity. However, during the seasons of planting spring or winter stem lettuce, especially in areas with heavy rainfall where conditions favor diseases like downy mildew, bacterial soft rot, and sclerotinia rot, well-draining soil should be selected. Crop rotation is advisable in soils prone to these diseases. When cultivating stem lettuce, it's beneficial to incorporate a substantial amount of barnyard manure and compost during plowing, especially for spring stem lettuce. If successive crops of other spring vegetables are planted in the same field the following spring, it's crucial to apply sufficient basal fertilizer in advance.

Based on the local topography and intercropping situation, ridges of 1.3–2.6 m in width are prepared. In seasons with heavy rainfall, it is advisable to create raised ridges to facilitate drainage. In colder regions, trench planting can be employed to protect the plants from cold winds. The field planting distances vary according to the varieties and seasons. For early-maturing varieties, a spacing of approximately 24 cm × 20 cm between rows and plants is suitable, while for medium-and late-maturing varieties, a spacing of around 33 cm × 27 cm is recommended. Stem lettuce seedlings are delicate; During field planting, it is important to handle them gently and retain the surrounding soil to avoid damaging the root system. Planting is best done on cloudy days or when soil humidity levels are suitable. After planting, timely watering is essential to enhance the chances of survival.

(3) Fertilization. Generally, stem lettuce requires 3–4 rounds of topdressing. After successful planting, it is necessary to apply fertilizer once to promote root system and leaf growth. During the rosette leaf formation period, when the stems start to enlarge, timely application of heavy fertilizer is essential to support stem growth. However, stem lettuce is sensitive to concentrated fertilizers, and the maximum concentration should not exceed 50% of regular organic manure. Late application of topdressing can lead to stem cracking and should be avoided. For spring stem lettuce overwintering, except for one round of topdressing after successful planting, no additional fertilization is needed during winter to prevent frost damage in cold regions. After the arrival of warm spring weather, timely fertilization is necessary to stimulate leaf and stem growth. For autumn stem lettuce, it is recommended to apply heavy topdressing half a month after field planting and light compost after 40 days.

(4) Pest and disease control. During the rainy seasons of spring and autumn, stem lettuce are prone to diseases such as downy mildew, sclerotinia rot, gray mold, leaf spot disease, and viral diseases. Common pests include aphids, thrips, and ground beetles. Downy mildew and sclerotinia rot are particularly harmful, significantly affecting yield. Prevention and control methods include ventilation and drainage to reduce air and soil humidity. Continuous cropping, overcrowding, and excessive watering should be avoided, and regular shallow hoeing to keep the soil surface dry is essential. It is recommended to remove diseased leaves, and preventively apply a 0.5% Bordeaux mixture before the occurrence of pests and diseases. In addition, it is necessary to remove infected plants promptly, clear away dead leaves, and burn them in a controlled manner.

(5) Harvesting. The standard for harvesting stem lettuce is when the heart leaves are even with the outer leaves, commonly referred to as a "flat mouth," or before the buds fully open. At this stage, the stems are fully enlarged and of high quality, tender, and crisp. Harvesting too late causes the flower stems to elongate, increasing fiber content, making the flesh hard, and sometimes even hollow, leading to reduced quality. Harvesting too early affects the overall yield.

2. Cultivation of leaf lettuce

Choose fertile or sandy loam soil with rich organic matter, good water retention, and loose texture, preferably loam with a pH value of around 6. Use newly harvested seeds in the current year to cultivate strong seedlings. Leaf lettuce cultivated in winter and spring, whether it's heading leaf lettuce or non-heading types, should adopt the seedling transplantation method. For leaf lettuce cultivated in summer and autumn, heading types usually adopt seedling transplantation method, while non-heading types can adopt either the direct sowing method or the seedling transplantation method.

Plow the soil 7–15 days before transplanting and apply sufficient basal fertilizer. Transplanting should be conducted when the seedlings have fully developed 5 true leaves. Employ raised bed cultivation, and determine the planting density based on the types and variety characteristics. Generally, for non-heading varieties, plant rows should be spaced at 20 cm × 15 cm, while for heading varieties, plant rows should be spaced at 40 cm × 30 cm.

After transplanting, watering should be performed together with phased topdressing and intertillage in the prophase when the surface of the soil looks dry, thus maintaining sufficient moisture of the soil, promoting root system expansion, and encouraging the growth of the bulbous leaves. In the metaphase and anaphase, to ensure that the basal leaves remain vibrant and the head leaves grow tightly and compactly, it is necessary to continuously and evenly water the plant. Stop watering before harvesting to facilitate post-harvest storage and transportation. Soilless cultivation is expected to be the future direction for leaf lettuce varieties that are mainly consumed raw. Soilless cultivation of leaf lettuce offers rapid growth and a short growth period. Leaf lettuce can be harvested 25–40 days after transplanting. It produces high commodity value, as well as high-yield, high-quality, and pollution-free produce, making it worthy of strong promotion and application.

V. Common Cultivation Issues and Their Prevention Strategies

1. Sudden growth

In the metaphase of growth, the stem may become slender, the leaf internodes may lengthen, leaves may become thin and small, with thick epidermis and little mesophyll, resulting in low culinary value. In this case, the plant is prone to bolting and flowering, and this phenomenon is commonly referred to as "sudden growth". There are roughly four main reasons that cause "sudden growth": ① Poor and droughty soil with insufficient water supply. ② Insufficient fertilizer supply leads to poor vegetative growth, especially when the tender stems are elongating and enlarging, inadequate fertilization may cause rapid upward growth of the stem, resulting in thin and elongated stems. ③ High temperatures and high respiration rates, resulting in increased nutrient consumption, and a low allocation of dry matter to edible parts. ④ Excessive watering. The solution is to choose fertile soil with rich organic matter, select suitable varieties and appropriate seasons, enhance water and fertilizer management, maintain soil moisture without waterlogging, and ensure an adequate supply of fertilizer when the tender stems are elongating and enlarging.

2. Cracking

In the anaphase of lettuce stem enlargement, the fleshy stem may longitudinally crack, reaching deep into the middle of the stem. The cracked parts are yellowish-brown and prone to rotting, which reduces their culinary value. The causes of cracking include: ① Variety of the lettuce; ② Uneven water and fertilizer supply, and erratic droughts and waterlogging, especially when the fleshy stem has matured and its exodermis has hardened. At this stage, excessive watering can lead to rapid expansion of the fleshy stem, while the epidermis cannot expand with it, causing cracking.

3. Premature bolting

In the cultivation process of lettuce, premature bolting is the phenomenon in which bolting and flowering happen before the fleshy stem becomes enlarged or fully enlarged, which is also known as early bolting. The causes for premature bolting include: inappropriate variety selection; poor timing of sowing; improper seedling management leading to either overgrown or elongated seedlings; excessive planting density; inadequate management of fertilizer and water (such as high temperatures, drought, water scarcity, excessive nitrogen, and insufficient potassium); pest and disease damage, among other factors. The prevention and control methods include selecting appropriate varieties, sowing at the right time, enhancing seedling management to cultivate robust seedlings, maintaining appropriate planting density, improving fertilizer and water management, and implementing pest and disease control. During the rosette leaf formation period and when the stem begins to enlarge, as well as when the leaves start to expand, foliar spray should be performed with 15–20 mg/L chlormequat or naphthaleneacetic acid.

「**Summary of Sub-context**」

This sub-context primarily introduces the basic knowledge of leafy vegetables and also provides information on the cultivation of lettuce. It's important to focus on the key points of lettuce cultivation.

「**Expanded Knowledge**」

The Therapeutic Value of Lettuce in Diet

Recent research has found that lettuce contains a type of hydroxylated aromatic hydrocarbon that can break down carcinogenic substances called nitrosamines found in food. This helps prevent the formation of cancer cells and has a certain preventive effect on digestive system cancers such as liver and stomach cancer. It can also alleviate the reactions of cancer patients to radiation therapy or chemotherapy.

Sub-context 2 Production of Celery

Celery, also called dry celery, or medicinal celery, is a biennial vegetable in the Apiaceae family known for its thick and tender petiole. It is native to the Mediterranean, Sweden, Algeria, Egypt, and swampy areas in Western Asia, including the Caucasus. Wild celery is distributed in these regions. Celery was first cultivated by ancient Greeks around 2,000 years ago. It was initially used for medicinal purposes and later cultivated as a pungent vegetable. It was selectively bred to develop the thick-petiole variety we know today. Celery was introduced to China from the Caucasus and has a cultivation history of over 2,000 years. It has gradually been bred into the slender-petiole variety. Celery has a wide range of adaptation and is currently cultivated in various regions across the country. Especially in recent years, with the continuous development of vegetable production and the increasing demand, the cultivation area of celery has been expanding throughout the country. It has become one of the main leafy vegetables consumed by urban and rural residents in China. Celery is rich in vitamins, minerals, and volatile aromatic oils, giving it a distinct flavor. It can stimulate appetite and has functions such as lowering blood pressure.

I. Biological Characteristics

1. Botanical features

(1) Roots: Celery has a shallow root system, mainly distributed in the soil layer of 10 to 20 cm, with a lateral spread of about 30 cm. As a result, its absorption area is limited, and it has relatively weak drought and waterlogging tolerance. However, the taproot can penetrate deep into the soil and store nutrients, becoming thicker. If the taproot is cut, many lateral roots can develop. Therefore, celery is suitable for seedling transplantation.

(2) Stems: During the growth period of vegetative growth, the stem remains short, but as it continues to grow, it can eventually bolt and produce flowering stems, which are upright and reach a height of 1–1.3 m.

(3) Leaves: The leaves are produced on short, compact stems and are unipinnate or bipinnate, typically divided into 2–3 pairs of small compound leaves. Each leaflet is ovate and three-lobed, with serrated margins. The common petioles are long and thick, constituting the main edible part. They can be 30–100 cm in length. Underneath the epidermis of the petioles, there is a well-developed thick collenchyma. Excellent varieties have less fiber content and high quality. Oil glands are distributed in the parenchyma cells near the vascular bundles, secreting volatile

oils that give off a distinct aroma. The cross-section of the stem is nearly round, semi-circular, or flattened. The diameter of the cross-section of petioles varies: 1–2 cm for Chinese celery, and 3–4 cm for Western celery. The inner side of the petiole has furrows, and the size of the pith cavity varies depending on the varieties. Petioles can be deep green, yellowish-green, or white in color. Deep green petioles are more difficult to soften, while yellowish-green ones tend to soften more easily. Under conditions of high temperature, drought, and insufficient nitrogen fertilization, the development of thick collenchyma and vascular bundles can lead to a decrease in quality. Unfavorable cultivation conditions often lead to the rupture of parenchyma cells, resulting in hollow and less substantial petioles, which can affect the quality.

(4) Flowers: It has umbel inflorescences with small, yellowish-white flowers. The corolla has 5 separate petals, and it is typically pollinated by insects through cross-pollination, although it can also undergo self-pollination. Celery sown in the fall will bolt and flower in the spring.

(5)Fruits and seeds: The fruit is a bilobed capsule, spherical in shape, containing 1–2 seeds. When it matures, it splits along the central seam. The seeds are brown and small, with a thousand-seed weight of 0.4 g.

2. Requirements for environmental conditions

(1) Temperature: Celery requires a cool and humid climate, with an optimal growth temperature range of 15–20°C. Growth is unfavorable and quality declines when the temperature exceeds 26°C. However, celery seedlings can tolerate high temperatures during the seedling stage, and young plants can withstand low temperatures as low as -7°C. They can safely overwinter in the middle and lower reaches of the Yangtze River basin. Seeds germinate at around 4°C, with the optimal germination temperature range being 15–20°C. Germination typically occurs within 7–10 days, but it is slower under high temperatures.

Celery is a vernalization plant, requiring low temperatures during the vernalization phase and long daylight hours during the photoperiod phase to trigger bolting and flowering. Celery can go through the vernalization phase in 10–20 days under temperature conditions of 2–5°C. Celery cannot go through the vernalization phase in germinating seeds and must reach a certain size as young seedlings to respond to low temperatures. Typically, Chinese celery requires 4–5 leaves, while Western celery needs 7–8 leaves or more to undergo the vernalization phase. Therefore, early spring sowing can lead to premature bolting, while celery sown in the fall must go through the winter to bolt and flower in the second year.

(2) Light: During the vegetative growth stage, celery does not have strict light requirements. However, after going through the vernalization phase, it must be under long daylight conditions to bolt and flower. During the seedling stage, it is beneficial to have abundant light, while in the anaphase of growth, gentle light is preferred to enhance both yield and quality. Seed germination requires weak light, and germination is poor in dark conditions. Western celery is a crop with relatively strict requirements for light intensity and should not be exposed to strong sunlight during its growth period.

(3) Water moisture: Celery requires moist soil and air conditions. Especially during the peak growth stage of vegetative growth when the surface is covered with white root hairs, adequate humidity is crucial. Otherwise, growth stagnates, and the mechanical tissues in the petioles develop, leading to a decrease in quality and yield.

(4)Soil and nutrients. Celery is suitable for well-drained, organic-rich, and loamy or clayey soils with good water retention and fertility. Celery has strict requirements for soil pH value, with an ideal range of 5.5–6.7. For healthy growth and development of celery, complete fertilizers are necessary. Nitrogen and phosphorus deficiency in the early stages have a significant impact on yield, while nitrogen and potassium are needed in the anaphase. Celery has a strong demand for boron. Lack of boron in the soil can lead to brown spots at the margins of celery leaves in the early stages, and in the anaphase, brown streaks and splitting of the vascular bundles in the petioles.

II. Types and Varieties

Based on the morphology of petioles, celery can be categorized into two types: Chinese celery and Western celery.

1. Chinese celery

The petioles are slender and reach a height of about 100 cm. It is categorized into green celery and white celery based on the colors of the petioles. Green celery plants are relatively larger, with larger green leaves and thicker petioles, measuring about 1.5 cm in diameter. They have a strong aroma, high yield, and good quality after softening. The petioles can be either solid or hollow. Solid celery has a very small petiole pith cavity, narrow and deep furrows, good quality, is less likely to bolt in the spring, has high yields, and is storage-resistant. Representative varieties include Beijing Solid Celery, Tianjin Baimiao Celery, Shandong Hengtai Celery, etc. Hollow celery has a larger petiole pith cavity and wider but shallower furrows, resulting in lower quality. It tends to bolt in the spring but has better heat resistance and is suitable for summer cultivation. Representative varieties include Fushan Celery, Xiaohuaye Celery, Zaoqin Celery, etc. White celery plants are relatively shorter and have smaller, pale green leaves with thinner petioles, measuring about 1.2 cm in diameter. The petioles are yellowish-white or white in color. They have a strong aroma, good quality, and tend to soften easily. Representative varieties include Guiyang White Celery, Kunming White Celery, Guangzhou White Celery, etc.

2. Western celery

Western celery has a larger plant size, reaching a height of 60–80 cm, with thick and wide petioles measuring about 2.4–3.3 cm in diameter. They are mostly solid and have a mild taste and crisp texture, but they are less heat-tolerant compared to Chinese celery. Individual plant weighs between 1 kg and 2 kg. It can be categorized into two major types based on the colors of the petioles: green-petiole and yellow-petiole celery. Green-petiole varieties have green, round, thick, and less fibrous petioles. They bolt late and exhibit strong resistance to stress and diseases. They have a late maturity period and are less prone to softening. Examples include Florida 683, Italian Winter Celery, Summer Celery, and American Celery. Yellow-petiole varieties have naturally golden-yellow petioles that do not undergo softening. The petioles are wide, thin, tender, and crisp but contain more fibers. They tend to be hollow early in their development, are sensitive to low temperatures, and bolt early.

III. Cultivation Season and Methods

For field cultivation of celery, it is advisable to schedule its vigorous growth period during cooler seasons. Therefore, autumn sowing is preferred, but it can also be grown in the spring. Since celery seedlings can withstand both high and low temperatures, autumn cultivation allows for earlier sowing to meet the demand for the September off-season market. Alternatively, late sowing can be done for winter and early spring harvests. In the Yangtze River basin, autumn sowing can be done from early July to early October, with crops sown in early July primarily harvested from September to October. Generally, early-maturing and heat-tolerant varieties are used, and shading and cooling measures should be taken during sowing. In some regions, such as Chengdu, sowing may begin as early as June. When sown in early August, harvesting can be done after January of the following year. When sown from September to early October, harvesting is typically completed in March to April of the following year, before bolting. Spring sowing is best done in March. Sowing too early may result in early bolting, while sowing too late can affect yield and quality.

In recent years, in the Yangtze River basin, there are several cultivation methods for growing celery in large-and medium-sized greenhouses.

(1) Large-and medium-sized greenhouse autumn (delayed) celery cultivation. Seedlings are sown from May to June, transplanted in mid- to late August, and when the weather turns cold in late October to early November and is no longer suitable for celery growth, large- and medium-sized greenhouses can be built with plastic film for insulation. Harvesting begins in early November.

(2) Large- and medium-sized greenhouse winter celery cultivation. Typically, seedlings are sown from early July to early August, transplanted from early September to early October, and the greenhouse film is covered from late October to early November. Harvesting is done from late December to mid-February of the following year, ensuring a supply for the Chinese New Year market.

(3) Large- and medium-sized greenhouse spring (overwinter) celery cultivation. Seedlings are sown from mid-August to mid-September, transplanted in late October

to early November, and when the temperature begins to drop in early November, the greenhouse film should be covered in a timely manner. Harvesting begins in mid- to late March of the following year, just in time for the spring market.

(4) Large- and medium-sized greenhouse summer celery cultivation. Seeds of summer celery cultivation are sown from the end of frost in spring to early to mid-May, transplanted in early June, and in late June, shade nets are used to reduce the sunlight inside the greenhouse. It is best to cover the shade net on top instead of on the edges, during sunny days instead of cloudy days, during daytime instead of nighttime, and cover the shade net in the prophase and remove it in the anaphase. Celery of this season can be harvested during the mild autumn months of August to September.

The cultivation of greenhouse celery can generally be divided into autumn-winter cultivation and spring cultivation. Autumn-winter cultivation can involve sowing in batches in late May or early June to September, with harvesting taking place from November to February for those who sowed in May or June, while those who sowed in September can harvest before the plants bolt in March or April of the following year. Spring cultivation can involve sowing after mid-December, with harvesting typically taking place before entering the hot season in June of the following year.

Soilless cultivation of celery often relies on substrate cultivation, with methods such as rock wool cultivation and substrate trough cultivation being commonly used.

IV. Cultivation Techniques

1. Sowing and seedling raising

Celery can be sown directly in the field or grown as seedlings for later transplanting. When sowing and raising seedlings in late summer to early autumn, it's advisable to choose a shaded location for the seedbed. Before sowing, it's important to deeply plow and expose the soil to the sun. Apply plenty of well-rotted compost and stable manure to maintain soil looseness, fertility, and moisture. Before sowing, soak the seeds in water at 55°C for 30 minutes. Then, wrap the seeds in damp gauze and conduct temperature stratification germination at 15–20°C. Celery seeds can be sown after 3–4 days (7–12 days for Western celery seeds); that is, the seeds can be sown when approximately 80% of the seeds have germinated.

Generally, the seeding rate range is 22.5–37.5 kg/hm^2, which can cover an area of 3–4.5 hm^2 for cultivation. After emergence, it is essential to enhance nutrient and water management, prevent damage from torrential rain, provide shade during the prophase to protect from strong sunlight, and pay attention to hardening off the seedlings in the anaphase. For those who sow seeds in September or spring, there is no need for seed soaking and germination induction, and the seeding rate is approximately 7.5 kg/hm^2.

2. Soil preparation and transplanting

Celery is suitable for well-drained, organic-rich, and soils with good water retention and fertility. Celery cannot be grown in low-lying areas. Proper planting density and soil preparation for celery are essential measures to achieve high yields and quality. After close planting or softening, the collenchyma of the celery petioles is less developed, and there is an increase in parenchyma cells, resulting in thicker and more petioles. For direct sowing, when the Chinese celery seedlings reach a height of about 3 cm, they should be thinned out. When the seedlings reach a height of around 12 cm and have 4–5 leaves, they are ready for transplanting or planting in their final location. In the southern regions, for celery sown in late summer to early autumn, the recommended row spacing for thinning should be about 6 cm between plants. For those sown later, the row spacing can be increased to about 12 cm when thinning the seedlings. When using this planting density, there is generally no need for soil softening.

If celery is cultivated by the softening cultivation method, the width of the furrow can be 1.3–1.6 m or 2.6–3.3 m, depending on the specific conditions in each area. Plant spacing should be approximately 6 cm. Generally, around 50 days after sowing, select the larger seedlings for transplanting. Planting should not be too shallow or too deep to avoid affecting root development and growth.

Softening in celery cultivation is generally done when the seedlings are about 30 cm tall, taking into account the dryness of the air, soil, and seedlings. Care should be taken to avoid injuring the plants or letting soil particles fall between the central leaves to prevent rot. Soil hilling is typically done after the cool autumn weather sets in. For early plantings, it is done 1–2 times, while for late plantings, it is done 3–4 times. Each time the soil is mounded, it should be done to a height that does not bury the central leaves. Generally, celery planted in the spring does not undergo soil mounding for softening.

Western celery can be transplanted when it has 8–9 true leaves as young seedlings. The planting density should be determined mainly based on the harvesting standards, considering the characteristics of the variety. If harvesting large-sized Western celery plants, 2 rows can be planted per ridge with a spacing of 25–30 cm between plants. If harvesting medium-sized Western celery, you can plant 3 rows per ridge with a spacing of 20–25 cm between plants. When transplanting, the seedlings should not be planted too deep, ensuring that the growing point of the seedlings is not buried in the soil.

3. Water management

After transplanting the Chinese celery in each season, water should be applied at regular intervals depending on the weather conditions until it ceases after soil hilling. The goal is to keep the soil slightly moist without causing whitening. Watering in the high-temperature season is best done in the morning and evening. In summary, the principle of water management is to water lightly and frequently to prevent excessive watering.

Western celery requires a large amount of water, and an adequate water supply is essential for high yields and quality. For those who cultivate celery in the summer and autumn, in the first week after transplanting, watering should be done once in the morning and once in the evening to promote the recovery of seedlings. One week later, the soil should still be kept moist, but it is advisable to use sprinkler or drip irrigation, and minimize the use of sprinkler or surface irrigation to reduce the occurrence of leaf spot disease. For those cultivating celery in the winter and spring, it is advisable to water around midday during the initial stage of transplanting, avoiding watering in the morning or evening.

4. Fertilization

Celery has a shallow root system, and due to the high planting density, in addition to providing sufficient base fertilizer, it's important to apply frequent but light topdressing. Continuously supply readily available nitrogen fertilizer along with phosphorus and potassium fertilizers. During the field growth period, it's important to apply topdressing fertilizer 2–3 times to promote healthy growth. When celery has grown 2–3 true leaves, well-rotted compost or diluted manure can be applied 1–2 times. The final topdressing should be done 5–7 days before hilling the soil. During the vigorous growth of celery, it is important to reapply topdressing, especially nitrogen and potassium fertilizers.

5. Pest and disease control

Celery is susceptible to diseases such as early blight, leaf spot, sclerotinia rot, viral diseases, soft rot, and black spot. The main pest affecting celery is aphids. Integrated pest management should be adopted with an emphasis on prevention. This includes maintaining field cleanliness, practicing crop rotation, and using seed disinfection methods. In the early stages of disease development, spray a 50% mancozeb wettable powder at a concentration of 500-fold dilution to prevent early blight; spray a 25% mancozeb wettable powder at a concentration of 300-fold dilution to prevent leaf blight and sclerotinia rot; spray a Bordeaux mixture at a ratio of 1 ∶ 0.5 (copper sulfate to hydrated lime) at a concentration of 160–200-fold dilution to prevent black rot; spray a dazomet zinc solution at a concentration of 500–600-fold dilution to prevent soft rot. Spray a 40% imidacloprid at a concentration of 1,000-fold dilution or a 50% cypermethrin solution at a concentration of 5,000-fold dilution to prevent aphids.

During the cultivation of celery, if environmental conditions are not suitable, there may be abnormal growth and development leading to physiological disorders. Planting in acidic soils deficient in calcium can lead to the occurrence of blackheart disease. To prevent this, it is advisable to apply an appropriate amount of lime to acidic soil before planting. Fertilization should involve a balanced combination of nitrogen, phosphorus, and potassium, with a careful avoidance of excessive nitrogen application. In the case of disease occurrence, foliar spray with 0.5% calcium chloride or calcium nitrate solution 2–3 times can be performed. Soil deficiency in boron can result in cracking of petioles. To prevent this, foliar spray with a 0.05% to 0.25% borax water solution can be performed. Celery etiolation and loss of green color are more likely to occur in high-calcium soils. If it first appears on new leaves, it indicates a manganese deficiency; If it appears on older leaves, it indicates a magnesium deficiency. The preventive measures include: In the early stages of the disease, targeted foliar sprays can be applied with 0.05%–0.1% manganese sulfate or 0.5% magnesium sulfate solution, once a week. The appearance of hollow petioles in solid-petiole celery varieties will reduce their market value. The occurrence of hollow petioles in celery is primarily due to its extended growth period, which leads to the natural aging of the leaves. This aging process starts from the outer leaves and progresses towards the inner ones, gradually resulting in hollow stems. Additionally, factors such as high temperatures, drought, insufficient irrigation, growth stagnation, delayed harvesting in the fall or winter, exposure to low temperatures due to delayed harvesting in the fall or winter cultivation, or rapid growth in high temperatures can also contribute to hollow petiole formation. To address this issue, it is essential to enhance irrigation management, amend sandy soils with organic fertilizers, and ensure timely harvesting.

6. Harvesting

For Chinese celery, the harvesting requirements may vary depending on factors such as the sowing time, cultivation methods, and specific varieties used. In general, except for early autumn sowing that may involve multiple harvests, sowing at any other time typically involves a single round of harvest. For late summer to early autumn or spring cultivation, the yield range can be 15,000–22,000 kg per hm^2. For winter harvesting, the yield can be as high as 45,000–52,500 kg per hm^2. After harvesting, celery can be stored using methods such as simulated planting storage, refrigeration, or controlled atmosphere storage.

To achieve high yield and maintain quality, it is essential to harvest celery promptly. During harvesting, cut the plants from the ground and then proceed with trimming. For domestic sales, typically, you only need to remove the outer 3–4 older leaves before selling. For exports, the requirements are relatively strict. Typically, you need to remove 4–5 outer leaves, cut off the leaf tips, and leave about 40 cm long petioles with a small amount of leaves. In recent years, soilless cultivation of celery has often relied on substrate cultivation, with methods such as rock wool cultivation and substrate trough cultivation being commonly used.

V. Common Cultivation Issues and Their Prevention Strategies

1. Uneven emergence and seedling deficiency after sowing

Under normal germination conditions, seedling deficiency and uneven emergence are mainly caused by improper water moisture management in the seedbed, leading to fluctuations between dry and wet conditions on the bed surface. The solution is to pretreat the seeds with a priming process, ensure the seedbed is well-watered, and maintain soil moisture.

2. Celery hollow petiole

This is a physiological aging phenomenon. The main reasons for this are poor and dry soil, insufficient funding, inadequate anaphase fertilization, frost damage to celery, and late harvesting, among others. Therefore, during the cultivation process, it is important to choose pure and high-quality solid petiole varieties. It is advisable to select non-sandy soil rich in organic matter, with good water retention and drainage in the field. Additionally, proper water management during the vigorous growth period should be emphasized, as well as maintaining soil moisture with a focus on readily available nitrogen fertilizer and supplementary potassium and boron fertilizers. Winter protection should also be considered, and timely harvesting is essential.

3. Premature bolting

Celery that does not meet harvest standards or even begins to bolt and flower during the seedling stage will lose its marketable characteristics. The main reasons for this are improper variety selection and either early spring sowing or late autumn sowing. The solution is to choose the appropriate variety and sowing time based on the climate conditions.

「**Summary of Sub-context**」

This sub-context mainly discusses the variety of selection and cultivation techniques for celery.

「**Expanded Knowledge**」

The Therapeutic Value of Celery in Diet

(1) Liver-soothing and blood pressure-lowering.

(2) Calming and nerve-soothing.

(3) Diuretic and anti-swelling.

(4) Anti-cancer and cancer prevention.

(5) Nourishing blood and tonifying deficiency.

(6) Celery is also an ideal green food for weight loss.

「**Reviewing and Thinking Questions**」

I. Explanation of Terms

1. Premature bolting.
2. Soil mounding.
3. Sudden growth.

II. Gap Filling

1. Lettuce can be classified based on the shapes of its leaves into________ and ________. Stem lettuce is commonly referred to as________ , leaf lettuce is commonly referred to as________.

2. The key techniques for high-yield and high-quality celery cultivation are ________. Celery is divided into________ and________ , which can be further categorized based on the degrees of petiole solidness into________and________ varieties.

3.________ deficiency often leads to cracking and hollowing of petioles, with initial symptoms including brown spots on the leaf margin.

4.________ is the principle for celery moisture management to prevent ________.

5. Prevention of________is common issues in leafy vegetables.

6. Lettuce can be classified into________and________based on the edible parts.

7. Lettuce is a________plant, and its seeds require________ conditions for germination.

8. In the cultivation of stem lettuce for its stems, the stems are more prone to becoming thin and elongated. The reasons for this are: ①________; ②________; ③________.

9. The harvesting standard for stem lettuce is that the________ and________ should be flat, commonly referred to as________.

10. A deficiency of________ in stem lettuce can lead to stunted growth, low yield, dark green leaves, and poor overall growth.

III. Choice Questions

1. The element that is needed in the largest quantity in celery cultivation is ( ).

A. Nitrogen B. Phosphorus

C. Potassium D. Boron

2. In the following ( ) group, there is a vegetable that does not belong to the leafy greens category.

A. Spinach, celtuce, lettuce B. Spinach, purslane, *Glebionis coronaria*

C. Celery, lettuce, amaranth D. Spinach, celery, kale

3. The type that does not belong to celery petiole color is ( ).

A. Green celery B. Yellow celery

C. White celery D. Purple celery

4. The plant growth regulator that can promote the growth of leafy vegetables is ( ).

A. Maleic hydrazide B. Gibberellin

C. Paclobutrazol D. Chlormequat

IV. Thinking and Answering

1. What are the characteristics of fertilization and water management for leafy vegetables?

2. What role do leafy vegetables play in achieving a year-round balanced vegetable supply?

3. Briefly describe the key technical points for greenhouse cultivation of leaf lettuce.

4. Analyze the reasons for the slender and weak growth of celtuce during cultivation and suggest measures to overcome this issue.

5. How to ensure that celery has an even germination during sowing?

6. Based on the factors contributing to celery yield, explain the main technical aspects of high-yield cultivation.

Learning Context 7 Production of Allium Vegetables

「**Learning Objectives in This Context**」

Master the main types and common characteristics of allium vegetables, the biological characteristics of the main types, and basic knowledge of allium vegetable production. Be able to choose the appropriate varieties of allium vegetables based on cultivation facilities and seasons. Be capable of creating production plans for allium vegetables and carrying out production correctly.

「**Analysis of Tasks in This Context**」

Master the technical aspects of land preparation, fertilization, ridging, sowing, planting, nutrient and water management, pest and disease control, harvesting, and post-harvest handling for Allium vegetables.

「**Introduction**」

Allium vegetables primarily include onions, garlic, chives, etc. They are important aromatic vegetables, and at the same time, various allium vegetables have antibacterial and health-promoting properties. At the same time, they are also an important component of China's export earnings from vegetables.

Overview

I. Types of Allium Vegetables

Allium vegetables have a long history of cultivation in China and come in various types. There are abundant local specialties, and commonly grown varieties across different regions, including Chinese chives, scallions, garlic, onions, shallots, and leeks (Fig. 7-1). These vegetables belong to the genus *Allium* in the Liliaceae family. They are biennial or perennial herbaceous plants. Due to their distinctive pungent aroma, they are also referred to as aromatic vegetables and are sometimes called bulb vegetables because they form bulbs. The edible parts of these vegetables are enlarged bulbs, pseudo stems, tender leaves, or flower stalks. *Allium* vegetables are rich in carbohydrates, proteins, minerals, various vitamins, and carotenes, making them highly nutritious. Additionally, the tissues also contain a white, oily, volatile substance called allicin, which can stimulate propylene sulfide, possess antibacterial and anti-inflammatory properties, and have significant medicinal value (such as extracting allicin from garlic and extracting Onionin A from onions).

Chinese chive　　Onion　　Garlic

Fig. 7-1　Different Types of Allium Vegetables

II. Biological Commonalities of Allium Vegetables

Allium vegetables mostly originated from the West Asian continental climate, and they share common growth characteristics: ① Morphological characteristics. They have shortened, disc-shaped stems, a fibrous root system that prefers moisture, drought-resistant leaf shapes, and storage organs in the form of bulbs or pseudo stems. ② Environmental requirements. They have a wide temperature range, preferring cool and temperate climates, and are cold-resistant. They require moderate light intensity. The above-ground growth period requires moist soil and low air humidity, while the bulb formation period requires relatively dry soil and air. They are long-day plants, belonging to the green plant vernalization type, and longer daylight hours and higher temperatures are conducive to bulb formation. ③ Reproduction method. They have the characteristics of tillering and various methods of reproduction.

III. The General Cultivation Principles of Allium Vegetables

Common characteristics of the cultivation of allium vegetables: ① Plants are low and compact, with upright leaves and small leaf area, suitable for dense planting or relay intercropping; ② Reproduction can occur through both sexual and asexual methods, with short seed viability, so freshly harvested seeds should be used for production; ③ Weeds should be promptly removed; ④ With a limited ability to absorb nutrients from the roots and high planting density, proper water and nutrient management is essential; ⑤ They share common pest and disease issues, so crop rotation should be practiced to avoid continuous cropping; ⑥ The root system can secrete antimicrobial substances, making them an ideal precursor or intercrop for other crops.

Sub-context 1 Production of Chinese Chive

Chinese chives, native to China, are perennial herbaceous plants in the *Allium* genus of the Amaryllidaceae family. They are known for their tender leaves, soft flower stems, and edible flowers. Chinese chives are cold-resistant and heat-tolerant, making them highly adaptable and suitable for cultivation in various regions across the country. Chinese chives are nutritionally rich, containing vitamins, minerals, sugars, proteins, and cellulose. They have a unique flavor that enhances appetite, making them a favorite among the general population. The seeds and leaves can also be used in medicine, with medicinal properties that include promoting digestion, invigorating the spirit, liver tonification, and kidney support.

I. Biological Characteristics

1. Botanical features

(1) Roots: They have fibrous root systems, with roots growing from the base of the rhizomes. The main root system is distributed within the 10–30 cm plow layer. Besides their absorbing function, they also serve as a storage organ, and can be divided into absorbing roots, semi-storage roots, and storage roots. In spring, they sprout absorbing roots and semi-storage roots, and these can produce 3–4 levels of lateral roots. In the autumn, short and thick storage roots develop, without any lateral roots. As the plant ages, tillering continues, and new roots continue to grow while old roots gradually wither, resulting in a constant turnover of roots. The location of root formation and the root system on the rhizome also move upward year by year, which is referred to as "root jumping". In production, it is necessary to continuously cultivate or apply soil and fertilize to prevent the rhizomes from being exposed, ensuring their normal growth. The lifespan of the roots is typically 1–2 years.

(2) Stems: The stem is divided into two types: nutritional stem and flower stem. The stem is developed from the embryo bud. In the case of 1–2-year-old Chinese chives, the nutritional stem is a shortened stem disc, with leaves growing at the top of the rhizome, and the base produces adventitious roots. As the plant ages, the nutritional stem continues to grow upward, connecting the sequentially generated tillers and stem discs into a branched structure, known as a rhizome. The rhizome is an important organ for storing nutrients. The leaf sheaths are fused to form a pseudostem, and the base of the pseudostem swells into a gourd-like shape, serving as an organ for nutrient storage. After entering the reproductive growth stage through vernalization, the terminal bud of the bulb differentiates into a floral bud. Under long daylight conditions, it produces a flowering stem known as a chive stem. At the top, it bears an umbel, and the tender stems are edible.

(3) Leaves: Chinese chives have clustered leaves, with each plant having 5–9 leaf blades. The leaves are composed of leaf blades and leaf sheaths. The leaf blades are flat and ribbon-like, serving as the primary assimilation and product organ. The leaf sheaths close to form a tubular pseudostem. The leaf surfaces are covered with wax powder, making them drought-resistant. The leaf's meristem zone is located at the base of the leaf sheath, and it can continue to grow after harvesting. The base of the leaf sheath swells, forming a gourd-shaped small bulb, which serves as a storage organ for nutrients.

(4) Flowers: Umbel with bisexual flowers. Before opening, they are enclosed by the involucre. Each inflorescence contains 20–30 flowers with white or pink corollas. These are bisexual flowers and rely on cross-pollination. Both tender flower stalks and flowers can be consumed. Chinese chives belong to the vernalization type of plants. Chinese chives sown in the same year rarely bolt without undergoing low-temperature vernalization in the autumn and winter. Chinese chives that are biennial or above usually start bolting from the end of summer to early autumn, with flowering occurring from early to mid-autumn.

(5)Fruits and seeds: Chinese chive fruits are capsules with a superior ovary composed of 3 chambers, each containing 2 seeds. When the fruits mature, the seeds burst out. The seeds are black, shield-shaped, with a surface covered in fine wrinkles, and the back side is convex while the belly side is concave. The thousand-seed weight is about 4 g and the seeds have a lifespan of 1-2 years (Fig. 7-2).

Fig. 7-2 Leaves, Flowers, and Seeds of the Chinese Chive

2. Requirements for environmental conditions

(1) Temperature: Chinese chives prefer cold and cool climates, with strong cold resistance and intolerance to high temperatures. The optimal temperature for

leaf growth is 12–24°C. When the temperature exceeds 25°C, plant growth slows, fiber increases, and quality decreases. At temperatures above 30°C, the leaves turn yellow and may even wither. Chinese chives are cold-tolerant vegetables, and their leaves can withstand temperatures as low as -5°C or even lower. The underground rhizomes are not subject to frost damage even at temperatures as low as -40°C. The following spring, when the temperature rises to 2–3°C, Chinese chives start to regreen and sprout new leaves.

(2) Light: Chinese chives are not very sensitive to light during their vegetative growth phase. Moderate light intensity is favorable for nutrient synthesis and accumulation.

(3) Water moisture: The root system of Chinese chives has a weak absorption capacity and prefers moisture. It requires the soil to be kept consistently moist to meet the needs of plant growth and development. If the soil lacks water, the mesophyll tissue often becomes coarse and tough, with increased fiber content, leading to a decrease in quality. Chinese chives have narrow, small leaves with a waxy coating, a thicker cuticle, and deep-set stomata. This reduces water evaporation, making them drought-resistant. They thrive with relatively low air humidity levels, ideally around 60%–70%. However, Chinese chives require higher soil humidity levels, ideally at 80%–95%.

(4)Soil and nutrients: Chinese chives have a strong adaptability to different types of soil, but they thrive in deep, organic-rich soil with good water retention properties. They also have a certain degree of tolerance to saline-alkali soils, and mature plants can grow normally in soils with a salt content of 0.25%.

Chinese chives prefer fertile soil and can tolerate rich soil. When cultivating them, it's important to provide sufficient basal fertilization. Additionally, during the spring and autumn seasons, topdressing should be applied in stages, mainly with nitrogen fertilizer, along with phosphorus and potassium fertilizers. This practice can help increase yield and improve quality. Chinese chives have varying fertilizer requirements depending on their growth stages. During the seedling stage, they have weak absorption ability and their nutritional needs are minimal, so it is best to apply small amounts of fertilizer regularly. In the peak vegetative growth stage, especially during spring and autumn harvesting, when nutrients accumulate, it's necessary to apply fertilizer in several stages with higher quantities. Annual seedlings require less fertilizer, while triennial to quadrennial Chinese chives need more fertilizer. Quinquennial Chinese chives, in order to extend their lifespan and slow down aging, should still receive fertilization, and this aspect should not be overlooked.

II. Types and Varieties

Chinese chives can be classified into four types based on the edible parts: root chives, leaf chives, flower chives, and flower-leaf chives. Currently, most varieties are of the flower-leaf chive type. In production, Chinese chives are generally categorized into two types based on leaf width: broad-leaf and narrow-leaf chives. Broad-leaf chives are primarily used.

1. Broad-leaf chives

Also known as "large-leaf variety" or "Malan chives", broad-leaf chives have broad, thick leaves that are light green or green. They have fewer fibers, high yields, and good quality, but their fragrance is relatively mild. They have poor upright growth and are prone to lodging. They are suitable for protected cultivation. Excellent varieties include Hanzhong Dongjiu, Henan 791, Tianjin Dahuangmiao, Beijing Dabaigen, Shouguang Majiu, Hangzhou Dongjiu, Jiangnan Mabianjiu, Chengdu Majiu, Guangzhou Dayejiu, etc.

2. Narrow-leaf chives

Also known as "small-leaf variety" or "Xian Jiu", narrow-leaf chives have slender, dark green leaves with more fiber, a strong fragrance, numerous tillers, thin and tall leaf sheaths, strong upright growth, and good resistance to both cold and heat. Excellent varieties include Beijing Tiesimiao, Baoding Honggen, Taiyuan Heijiu, Zhucheng Dajingou, etc.

III. Cultivation Season and Methods

Chinese chives are cold-resistant, shade-tolerant, and highly adaptable, making them suitable for year-round outdoor cultivation in regions south of the Yangtze River. In areas north of the Yangtze River, Chinese chives' above-ground parts wither during the winter, and their root rhizomes go dormant, protected by the soil. Chinese chives can be sown in both spring and autumn. In winter and early spring, you can produce green chives in protected structures such as sunlight greenhouses, plastic tunnels, and sunbeds.

IV. Cultivation Techniques

1. Reproduction method

Since the growth of plants propagated by division is not as strong as those propagated from seeds, most of the production is done through seed propagation. Seeds are sown separately in spring or autumn for seedling cultivation, and they are transplanted in the following autumn or the spring of the following year. The seeds must be fresh seeds harvested from the previous year or the current year. The seeding rate on a nursery bed is 3–5 kg per mu, and the resulting seedlings can be used for cultivation over an area approximately 10 times larger.

2. Field planting

(1) Transplanting period: When the plant height reaches 20–25 cm or when seedlings are crowded, they should be transplanted promptly. Generally, transplanting can be done 50–60 days after emergence. Transplanting should be done no later than late July. Otherwise, it falls into the high-temperature and rainy season in the South during July and August, which is not conducive to the survival of young seedlings.

(2) Land preparation, fertilization, and ridge formation: Chinese chives require substantial fertilization and have a long growth period. Therefore, it's necessary to apply a heavy dose of basal fertilizer before transplanting. After the previous crop is harvested, apply 4–5 tons of well-rotted, high-quality farmyard manure per mu. Deep plow to a depth of 25–30 cm, finely rake, and form ridges that are 1.6–2 cm wide and 20–25 m long.

(3) Reasonable close planting: Dig furrows with a row spacing of 50 cm and plant clusters at intervals of 17–20 cm. Place 4–5 seedlings per cluster.

(4) Transplanting method: Before transplanting, thoroughly water the seedbed to facilitate seedling removal. When raising seedlings, shake off the old soil, and

for excessively long roots, trim the tips, leaving only 2–3 cm. At the same time, to improve the survival rate, trimming a section from the tips of the leaves is necessary to reduce leaf surface evaporation. Chive seedlings should be transported shortly after lifting and should not be stored for an extended period. The planting depth should be such that the leaf sheath is buried in the soil, typically around 3–4 cm deep. Planting too deep can lead to weak growth while planting too shallow can result in rapid root jumping. The soil around the chive clumps should be compacted, and then water should be applied to ensure good contact between the roots and the soil.

3. Field management

After transplanting in the autumn, the leaves grow rapidly, and the chives have strong tillering ability. At this time, it is important to intensify fertilization and irrigation to meet the requirements for growth and tillering. Before severe cold weather sets in, apply one heavy fertilization to promote growth, allowing more nutrients to accumulate in the underground rhizomes to meet the needs of sprouting and growth in the coming spring. Generally, the green chives should not be harvested at this time. After 2–3 years, harvest multiple times each year, and after each harvest, apply heavy fertilization to promote leaf growth and tillering. Furthermore, after each harvest, it is necessary to nurture the plants for a period, allowing them to recover and resume growth before the next harvest. This approach helps maintain vigorous growth and prevents premature aging. The specific length of time between harvests is determined based on the growth status and temperatures. After reaching the harvest age, it is essential to pay attention to the growth of underground rhizomes and accumulate more nutrients. Furthermore, after years of growth, tillering, and root jumping, the plants become crowded, and new tillers become smaller, leading to reduced yields. At this point, it is necessary to promptly remove old roots and leaves, amend the soil, and fertilize to promote recovery and growth.

In the hot and humid summer of the southern regions, it is not suitable for the growth of Chinese chive leaves. During this period, do not harvest to protect the plants through the summer. This way, after the cool autumn arrives, they can resume their growth. In regions south of the Yangtze River, although the aboveground parts of Chinese chives can overwinter outdoors, they are not entirely free from frost damage in areas on both sides of the Yangtze River. Varieties with weaker cold resistance can be covered with straw if necessary.

Soil mounding is an important part of the field management and is also related to the phenomenon of "root jumping". The more you harvest, the greater the distance the "root jumping" phenomenon occurs. Therefore, after each harvest, you should promptly mound the soil by 3–4 cm. Typically, this is done twice a year. Each time you harvest, it should be about 2–3 cm higher than the previous harvest.

4. Pest and disease control

Chemical weed control is very effective in Chinese chive fields. According to the experience in the outskirts of Beijing, the use of herbicidal powder in Chinese chive fields is applied after sowing, with a usage rate of 15 kg per hm^2. To avoid contact with the leaves, mix herbicidal granules with fine soil before application.

The rhizome and lower leaf sheaths of Chinese chives with deteriorated roots will grow in the soil for a long time, making them susceptible to various diseases and pests. The primary underground pests include Chinese chive root maggots and thrips. To prevent these pests, sugar and vinegar solution can be used to attract and kill adult insects before planting or before the pests become a problem. Pesticides like cyhalothrin, pyrethrum emulsion, or dimethoate, and quinalphos emulsion can also be used to control them, but highly toxic pesticides with high residues should be avoided. Root rot is due to the damage caused by the fusarium wilt of Chinese chives. This type of pathogen is more likely to occur in poorly-drained soils, especially during the hot summer season, after heavy rainfall, and when exposed to intense sunlight. The main prevention and control methods are to improve drainage by digging trenches and to cultivate strong seedlings.

5. Harvesting

In the southern China, except during the hot summer, green chives can be harvested almost year-round. In the Yangtze River basin, green chives are typically harvested 2–4 times from spring to summer. During harvesting, it is important to leave about 3–5 cm of the base of the leaf sheath. If the cut is green, it is too shallow; If the color is white, it's too deep. A yellowish cut is preferred. This helps prevent damage to the leaf sheath's meristem and young buds, preserving the yield for the next harvest. Between July and August, when temperatures are high and growth is slow, it's common to refrain from harvesting green chives and instead focus on collecting chive stems. After entering autumn, you can also harvest green chives once or twice, or choose not to harvest them, focusing more on root development. In winter, it is common to use soil mounding to soften the soil and then harvest yellow chives. The general annual yield is around 50 t/hm^2.

The quality of Chinese chives is not only related to the characteristics of the variety but also to temperature, light, moisture, and soil nutrition. The quality is generally better when harvested in the spring and autumn, especially after rain, as the leaves grow quickly and are tender. During the summer, in the hot season, the leaf growth is slow, and the fiber content is high, while the sugar content is low, resulting in lower yields. Winter-grown yellow chives, cultivated by softening cultivation method, have less fiber and more sugar content, making them a delicacy among vegetables.

6. Key points in facility cultivation

The cultivation of Chinese chives in a controlled environment can be done through various methods, including windbreak early cultivation, covered chive cultivation, and stacked chive cultivation. Using large- or medium-sized greenhouses for Chinese chive production during the winter and spring seasons has become one of the most common methods of controlled environment cultivation in China. The use of plastic film covering increases the temperature inside the greenhouse, making it possible to produce green chives during the winter and early spring, when they would not naturally be available. During the cultivation of Chinese chives in a controlled environment, several important considerations must be taken into account.

(1) Start of the covering. Chinese chives have a certain dormancy period, and covering them too early can lead to a short dormancy period, resulting in reduced yields and lower quality during the later harvest period.

(2) Temperature management. The temperature inside the greenhouse is easily

influenced by light, leading to significant fluctuations. The optimum temperature for the growth of Chinese chive leaves is 18–20°C. If it drops below 5°C, growth will stop. If it goes above 25°C, it can lead to symptoms like leaf apex curling, scorching, and wilting.

(3) Humidity management. When humidity is high, Chinese chives are prone to rotting. Therefore, it is necessary to take measures such as ventilation to reduce the humidity in the greenhouse and improve the yields.

(4) Harvesting frequency. To ensure increased production of green chives during the winter and spring, it is advisable to reduce the frequency of harvesting in the autumn. This allows more nutrients to be stored in the robust stems and roots for growth during the winter and spring seasons.

(5) Harvesting period. Greenhouse-grown Chinese chives are typically sown in early spring, transplanted around June, and their covering begins shortly after the last frost. In a sense, it is off-season cultivation, with relatively lower energy accumulation in the plant. Therefore, typically, after 2–3 years of planting and harvesting, the plants should be renewed.

V. Modern Cultivation Techniques for Yellow Chives

Softening cultivation is a method of growing certain vegetables in a dark (or low-light) and suitable temperature and humidity environment during a specific growth stage. This method results in plants with lower chlorophyll content, a yellowish-white appearance, tender tissues, unique flavor, and high commercial value. A cultivation method using black plastic mulch cover will be introduced here.

The cultivation of yellow chives aims to provide the ideal conditions of temperature, humidity, and darkness for the chive roots. By utilizing the nutrients stored within themselves, they grow without the formation of chlorophyll. This results in the production of tender, flavorful, pale yellow chive plants known as yellow chives. They have minimal cellulose content and a soft, tender texture. Generally, most Chinese chive varieties can be used for producing yellow chives, but it's best to select a specialized variety for yellow chive production, such as the Dun Yellow Chives. The sowing time is from late March to early June, and in warm southern regions, it can also be sown in late autumn, from September to October. The seeding rate is 1.5–2 kg per mu, and the seedlings can be transplanted when they have 4–6 leaves. In the spring and summer seasons, covering the Chinese chives with black plastic film can provide the necessary softening cultivation conditions for forced growth without light, maintaining warmth and moisture, and thus producing tender yellow chives. The specific cultivation techniques will be described as follows:

(1) Cultivation of rootstock. Generally, robust biennial rootstocks are chosen. In open fields, only one or two cuts of green chives are harvested from the rootstock, allowing the rootstock to accumulate more nutrients, which is beneficial for increasing the yields of yellow chives. Before covering, when the green chives are not yet harvested, apply well-rotted human manure once and add a small amount of urea as base fertilizer to enrich the rootstock.

(2) The covering method typically involves using a row cover with a tight-fitting black plastic film on top. Insert light-blocking boards at both ends of the ridge to facilitate ventilation and shading. Additionally, to regulate the temperature, prepare straw mats. The straw mats shall ideally be 2.5 m in length and 1 m in width. All of these preparations should be completed by mid-October. Before covering with plastic film, clear the dried leaves and weeds in the chive ridge. Use a chive hoe to remove the above-ground green parts that have not been harvested. After the wounds have healed, water thoroughly once, and then cover 2–3 days later. During the winter and spring seasons, when covering, attention should be given to temperature reduction and ventilation management. At the beginning of the covering period, insert shading boards at the two ends of the ridge' s ventilation openings to prevent direct sunlight from entering the beds. To prevent high temperature and humidity, and avoid leaf rot, a tile pipe should be buried every 3 m near the ground on both sides of the greenhouse to enhance ventilation and air exchange. During the first growth cycle of yellow chives, which occurs in the cold winter season, the greenhouse temperature is relatively low. To improve yellow chive growth, straw mats can be used to cover the greenhouse roof. These mats should be removed during the day and placed back in the evening to raise the

temperature inside the greenhouse. As the temperature gradually rises during the second growth cycle and humidity increases inside the greenhouse, it's advisable to increase the density of the tile pipes on both sides of the greenhouse. You can add one additional tile pipe between the existing tile pipes at the sides, and extend the duration of opening the tile pipes at both ends to facilitate better airflow. During the strong midday sunlight, straw mats can be used to lower the temperature inside the greenhouse, maintaining it around 20°C with a humidity of 60%–70%. This will provide the optimal conditions for vigorous growth of the yellow chives. Cultivating yellow chives under black plastic-covered greenhouses can shorten the growth period and increase the number of harvests. If timely topdressing and watering are provided during the harvesting period to prevent plants from depleting their nutrients prematurely, higher yields can be achieved.

VI. Common Cultivation Issues and Prevention Strategies

1. Seed issues

(1) Contamination of scallion seeds in the seedlot. Primary reasons: Due to the low yield and high price of Chinese chive seeds compared to the high yield and low price of scallion seeds, and similar appearance between these two types of seeds, many unscrupulous seed sellers mix scallion seeds into Chinese chive seeds to make a significant profit.

Prevention and control measures: Purchase seeds from reputable seed stores. When buying seeds, carefully observe and distinguish between them. Chive seeds typically have a shield-shaped and slightly flattened appearance with many fine wrinkles and no concave navel; Scallion seeds are shield-shaped with some angular flatness, irregular wrinkles, and a slightly shallow concave navel. Chinese chive seeds are slightly larger, with a thousand-seed weight of around 4.15 g; Scallion seeds are a bit smaller, with a thousand-seed weight of around 2.6 g.

(2) Low or zero germination rate. Primary reasons: seed aging, poor weather during seed production, and inadequate management during seed harvesting.

Prevention and control measures: Use fresh seeds only; improve management during seed production, and take preventive measures against adverse weather conditions during seed harvesting; improve management after seed production.

2. Plant degeneration

After several years of Chinese chive cultivation, the plants' tillering ability decreases, and their growth rate slows down. Primary reasons: inappropriate varieties; excessive harvesting; and insufficient management of fertilizer and water.

Prevention and control measures: Choose suitable high-quality varieties, control the number of harvests, and enhance fertilizer and water management.

3. Yellowing and drying of leaf apexes

Primary reasons: improper temperature and humidity management, soil acidification, poor ventilation in greenhouse cultivation, pesticide damage (excessive pesticide concentration), fertilizer damage (excessive application of nitrate nitrogen), and variety characteristics.

Prevention and control measures: Choose appropriate varieties, improve temperature and humidity management, amend the soil, control pesticide concentration, regulate the quantity of nitrate nitrogen fertilizer, apply micronutrients to supplement trace elements, and ensure timely ventilation in greenhouse cultivation.

「**Summary of Sub-context**」

This sub-context primarily covers the methods of propagating and planting Chinese chives, the timing and quantity of planting, and field management and harvesting techniques.

「**Expanded Knowledge**」

The Medicinal Functions of Yellow Chives

Yellow chives are considered a warming and yang-tonifying herb. They contain a significant amount of zinc, which can nourish the liver and kidneys. Therefore, they are referred to as the "yang restoring herb" in traditional herbal medicine. In addition, yellow chives contain volatile essential oils and sulfur-containing compounds, which have the ability to stimulate appetite and reduce blood lipids. They have some therapeutic effects on conditions like hypertension, coronary heart disease, and high blood lipids. Sulfur-containing compounds also have certain antibacterial and anti-inflammatory effects. Yellow chives contain a significant amount of dietary fiber, which promotes gastrointestinal motility and can effectively prevent chronic constipation and colon cancer. These dietary fibers can also wrap around impurities in the digestive tract and eliminate them from the body with the feces. Additionally, yellow chives have the effects of warming the stomach, promoting digestion, dispersing blood stasis, detoxifying, and providing warmth.

Sub-context 2 Production of Garlic

Garlic is a cultivated species in the *Allium* genus of the Liliaceae family. It is an annual or biennial herb characterized by bulbil-forming bulbs. It is originally from southern Europe and Central Asia, and it has been introduced to China for over 2,000 years. It is widely distributed in China, and the cultivation area and production are among the highest in the world. Garlic is used for its bulbs (including multi cloves and single clove), garlic scapes (flower stalks), and seedlings (green garlic or garlic sprouts). In the southern regions, the main focus is on harvesting garlic sprouts and garlic scapes, while in the northern regions, the primary harvest is garlic bulbs. Garlic is rich in proteins, carbohydrates, vitamins, and minerals such as phosphorus, iron, and magnesium. It also contains allicin, which gives it its distinctive spicy flavor. Garlic is not only used for seasoning and enhancing appetite but also possesses antimicrobial and antibacterial properties. Garlic is rich in nutrients with a unique flavor and has a wide range of uses. It can be stored for a long time and is known for its medicinal and health benefits, including its antibacterial, antifungal, and detoxifying properties. Garlic is believed to alleviate symptoms and contribute to the treatment of conditions like high cholesterol, diabetes, heart disease, and various types of cancer, including those affecting the stomach, intestines, liver, lungs, and breasts. Its medicinal value is highly regarded and beloved by people worldwide. Garlic can be used in cooking, and processed into various pickled products, condiments, and garlic powder, and it can also be used to extract garlic oil (with the main component being allicin). Therefore, it finds wide applications in daily life, medicine, the chemical industry, and the food industry.

I. Biological Characteristics

1. Morphological characteristics in botany

(1) Roots: Cord-like fibrous root system, originating at the base of the short stem, includes primary roots, secondary roots, and adventitious roots. From the base of the back and ventral side of the clove, root primordia are first formed. The roots that extend from these primordia are called primary roots. The roots growing around the "stem disk" at the base of the ventral side are secondary roots. The second batch of new roots that grow around the time of bulb swelling is known as adventitious roots. Garlic is a shallow-rooted crop with most of its root system concentrated on the outer side of the cloves, while the inner side has fewer roots. The primary root system is concentrated within the soil layer of 5–25 cm and has a horizontal diameter of about 30 cm. It has very few root hairs, weak absorption capacity, and has a preference for moisture, tolerance to fertility, and an aversion to drought.

(2) Stems: The stem of garlic plants is underground, and during the vegetative growth stage, it has a short and irregularly shaped structure, known as the "stem disc". The base and edges of the stem disc develop roots, while the upper part contains the primordial body of leaves and buds. The terminal bud is located in the middle of the upper part of the stem disc and is surrounded by layers of leaf sheaths. The stem disc supports pseudo stems, garlic scapes, and garlic bulbs, while also serving a transport function. During the reproductive growth stage, the terminal bud differentiates into floral buds, which later develop into flower stalks, also known as

garlic flowers and garlic scapes. At the same time, the base of the inner leaf sheath begins to form lateral buds, gradually developing into bulbils.

(3) Leaves: The leaves consist of leaf blades and leaf sheaths. The leaf blades are flat, lanceolate, green, and covered with wax powder on the leaf surface, which can reduce leaf surface evaporation and make it drought-resistant. The leaves are more erect, and the leaf area is small. The leaf sheaths are cylindrical, surrounding the stem disc and growing in layers, forming a pseudostem. The leaves are arranged in an alternate fashion, symmetrically, with their orientation perpendicular to the line connecting the back and front of the garlic cloves. The number of leaves varies depending on the variety, with more leaves resulting in a thicker pseudostem. When the bulbs enlarge, the nutrients in the leaves are transported to the bulbils. As the bulbs mature, nutrients in the lower parts of the outer leaf sheaths are transferred to the garlic cloves. The leaves gradually wither and then shrink into a membranous covering for the bulbs, providing protection.

(4) Flower stems and aerial bulbs: The garlic flower stem, known as the garlic scape, typically grows to a length of 60–70 cm. It is cylindrical and solid. At the top of the flower stem, it bears an involucre that encases the inflorescence. Inside the involucre, you can find multiple aerial bulbs and underdeveloped purple flowers, which do not produce seeds. Flowers and bulbs often coexist on the same stem. When the growth of the small bulbs inhibits the development of the flowers, the flowers typically wither prematurely. The aerial bulbs can be used for propagation through planting.

(5) Bulbs: Also known as garlic heads, they consist of bulbils, leaf sheaths, and short rhizomes. They are a collective structure of bulbils and the main organ of garlic. The shapes of the bulbs vary with different varieties and can be round, oval, or conical. Bulbils are often semilunar in shape, shorter in purple-skinned garlic varieties, longer in white-skinned garlic varieties, and single-bulb garlic is spherical, with a structure similar to that of typical bulbils.

2. Requirements for environmental conditions

(1) Temperature: Garlic prefers a cool climate, with its optimal growth temperature range being 12–25°C. After a period of dormancy, garlic cloves can

sprout at 3–5°C. Sprouting accelerates significantly at temperatures above 12°C, with the most favorable sprouting temperature being around 22°C. The optimal temperature for seedling growth is 14–20°C. During the period of elongation of garlic scapes and bulb swelling, the suitable temperature is around 15–20°C. In the later anaphase of growth, the optimal temperature is approximately 25°C. When the temperature exceeds 26°C, the plant growth slows down, leaves turn yellow, and the above-ground parts gradually wither. Bulb development stops, and the plant enters a dormant period. Garlic belongs to the vernalization type of plants. Generally, from the sprouting stage to the seedling stage, exposure to temperatures between 0°C and 4°C for 30–40 days helps it pass through the vernalization phase.

(2) Light: Both garlic bolting and bulb formation require the induction of long daylight hours. The critical lengths of long daylight hours vary depending on the garlic varieties. Different ecological types of garlic varieties respond differently to the bulb length of daylight hours in the development and formation of bulbs. Garlic of low-latitude types requires less cold and can form bulbs even under short daylight hours (8–10 hours) with increasing temperatures, resulting in early maturity. High-latitude types, need a certain duration of cold (3 months below 5°C) and long daylight hours (more than 14 hours) to develop bulbs, making them mid- to late-maturing. Therefore, when inducing garlic between different latitudes, it is important to consider the varying light requirements for bulb formation. Insufficient hours of light will result in the growth of garlic leaves but not in bolting or bulb formation.

(3) Water moisture: Garlic leaves belong to a drought-tolerant ecological type, but its shallow root system has a weak water absorption capacity. Therefore, it prefers moist conditions and is sensitive to drought, making it demand a higher level of soil moisture. During the sprouting period, higher soil humidity is required to facilitate root emergence and budding. In the prophase of seedling growth, soil humidity should not be excessive to prevent bulb rot. During the declining period of the mother bulb, soil humidity should be increased to prevent soil from drying out and promote plant growth, reducing the occurrence of “yellow tips”. During the period of garlic scape elongation and bulb

enlargement, which is the peak growth stage, garlic requires the most water, so keep the soil consistently moist during this time. As the harvesting period of the garlic bulbs approaches, reduce watering to lower soil humidity levels, which helps with bulb maturity and shelf life.

(4) Soil and nutrients: Garlic has a broad soil adaptability but possesses a weak and small root system. It thrives in deep, well-draining, loose, slightly acidic loamy soils rich in humus. It is not suitable for cultivation in nutrient-poor, low-organic matter, alkaline soils, or in areas where soil alkalinity increases in early spring. The optimal soil pH value for garlic is in the range of 5.5–6.0. If the soil becomes too acidic, the root tips may thicken, causing growth to halt. Overly alkaline soil can lead to the rotting of the cloves, and increased production of small bulbs, and single-clove garlic, which can reduce yields.

II. Types and Varieties

Garlic is generally categorized into two major types based on the colors of the skin: purple-skinned garlic and white-skinned garlic. Variety names are generally named according to the place of origin.

1. Purple-skinned garlic

The garlic bulb has a light purplish-red skin, so it is also called red-skinned garlic. Purple-skinned garlic has weaker cold resistance, a shorter growth period, fewer but larger garlic cloves, and a strong pungent flavor, and it produces large garlic scapes. It is often an early-maturing variety, suitable for spring planting, and it has excellent quality. It is well-suited for growing garlic greens, garlic scapes, and garlic bulbs. Cultivated varieties vary by regions. In Chongqing, the main cultivated purple-skinned garlic varieties include Tongzi Garlic, Siyue Garlic, and Ruanyezi Garlic, in Yunnan, there is Tonghai Garlic; in Chengdu, there is Ershuizao Garlic.

2. White-skinned garlic

White-skinned garlic has white or grayish-white outer and inner skin. Most white-skinned garlic varieties are winter-hardy and cold-resistant, and they tend to have smaller bulbs with a milder spicy flavor. They are suitable for growing as bulbs or garlic scapes. Cultivated varieties vary by regions. In Chongqing, some of the main white-skinned garlic varieties that are grown include Guizhou White Garlic and Wutai Garlic.

III. Cultivation Season and Methods

The planting period for garlic varies depending on market demand, variety characteristics, and intended use. It is generally categorized into two main types: spring planting for summer harvesting, and autumn planting for spring harvesting. In southern regions, garlic is generally grown in open fields, with a focus on autumn planting. In the Yangtze River basin, autumn planting is prevalent, typically conducted from early August to early October. Garlic cloves are planted during this period, and the young seedlings can overwinter in the field. With the influence of low winter temperatures, they pass through the vernalization phase. In these regions, garlic starts to bolt from March to May and is typically harvested from April to June. In North China, both autumn and spring planting are feasible. However, in Northeastern provinces, garlic is typically only planted in spring.

IV. Cultivation Techniques

1. Sowing

Garlic is cultivated by directly sowing individual garlic cloves. Before sowing, prepare the soil by plowing and applying fertilizer. Avoid continuous cropping, and use fertile and well-draining sandy loam, which is suitable for garlic cultivation.

(1) Selection of garlic cloves and pre-sowing treatment. The garlic cloves used for planting should be selected and graded before sowing. Strict selection should begin in the field before harvesting, especially to remove plants infected with viruses. After harvesting, select large, round and intact garlic bulbs that have the characteristic features of the variety.

The size of garlic cloves used for planting is positively correlated with future yield. Larger garlic cloves contain more nutrients, are less likely to be infected with viruses, and result in faster growth during the seedling stage. This leads to higher future yields of garlic bulbs and thicker garlic scapes. The use of virus-free seedlings produced through apical meristem culture not only results in high yields but also leads to large garlic bulbs. It has played a crucial role in production.

In production, to break dormancy and promote germination, you can remove the garlic skin before sowing, or soak the garlic cloves in water for 1–2 days before sowing. This aids in water absorption and gas exchange. Additionally, storing garlic cloves at 0–4°C (which can be done using cold storage or ice blocks in production) for one month can significantly accelerate germination. These methods are particularly suitable for cultivating green garlic, allowing for earlier planting and supply.

(2) Sowing period. In various southern regions of China, garlic is typically sown in the autumn after its dormancy. For garlic cultivation aimed at producing garlic bulbs, the sowing is done relatively later, mostly in late September. The bulbs are harvested in the following year, typically from early to mid-June. As for garlic cultivation aimed at producing green garlic, the planting is done earlier. It can be sown from July to August, and harvesting of green garlic (meaning the garlic leaves) typically starts from September to October of the same year. Regardless of whether it is cultivated for garlic bulbs or green garlic, the sowing period doesn't have the same strict concern for premature bolting as onions do. This is a fundamental difference between garlic and onions. The reasons include: ① The conditions required for garlic bulb and flower bud differentiation (a period of low-temperature induction) are consistent and simultaneous. It is not possible to form bulbs without bolting. ② Garlic scapes are also consumed as an edible part. ③ If the garlic scapes are harvested promptly, it will not affect the bulb yield.

2. Cultivation methods and planting density

Planting density and seeding quantity. The planting method is to insert the garlic cloves into the soil with their tips slightly exposed, and they should not be planted too deeply. The spacing between rows when harvesting garlic bulbs is (15–20) cm × (10–13) cm, or even denser. The seeding quantity per hectare is 750–2,000 kg. The wider the row spacing, the larger the individual bulb weight, but it's more prone to developing irregular bulbs. When aiming to harvest green garlic, the planting is denser with row spacing of 12 cm × (4–7) cm.

3. Chemical weed control

Timely and efficient weed removal is a crucial measure for high-quality and high-yield garlic cultivation. Chemical weed control is commonly used, and you can choose options like 50% Oxyfluorfen or a combination of Oxyfluorfen and Quizalofop-p-ethyl for pre-emergence treatment after sowing, or spray the stem and leaves when the garlic plants have 2–3 leaves in one heart stage.

4. Field management

When the seedlings emerge from the soil to a height of 3–6 cm, topdressing should be applied, primarily using nitrogen fertilizer. During the growth of green garlic, from August/September to November/December, it is advisable to apply topdressing 2–3 times to promote the growth of the above-ground part. The greater the growth of the above-ground part, the higher the yields. If the goal is to harvest garlic bulbs, in addition to applying topdressing during the seedling stage, one more round of fertilization should be done before the winter. The second year, after the arrival of spring and warm weather, comes the period of vigorous growth for garlic plants. At this time, one heavy application of fertilizer should be applied to promote the subsequent bulb development. Once the garlic bulbs start to swell, it is not advisable to apply excessive or concentrated fertilizer to avoid causing bulb rot. Topdressing with organic manure, especially composted human and animal waste, is essential to promote bulb development. To achieve high yields, it is crucial to maintain a substantial leaf area during the bulb formation stage.

Intertillage for weed control is particularly important after the emergence of garlic seedlings. When the seedlings are 10–15 cm tall, you can perform deeper intertillage. When they reach a height of 30 cm or more, the intertillage should be shallower. In general, intertillage and watering are often combined with fertilization. However, by the second year from April to May, when the bulbs have started to swell, and with increased rainfall, attention should be paid to drainage to prevent garlic clove splitting. If too much nitrogen fertilizer is applied, there is a possibility of new garlic cloves sprouting and producing additional leaves, which can affect the quality and storage characteristics of the bulbs.

5. Pest and disease control

The main diseases include leaf blight, purple spot disease, white rot, and rust disease. Leaf blight can be treated with 10% difenoconazole, 25% prochloraz, and 25% methasulfocarb; purple spot disease can be controlled using 3%/2% polyoxin, 41.5% prochloraz; white rot can be managed with 50% carbendazim, 20% tolclofos-methyl, 75% Suanyeqing and 50% iprodione; rust disease can be treated with Taigao, Bailitong, and 25% propiconazole.

The main pests include Asiatic onion leafminer, onion thrips, *Liriomyza huidobrensis*, and thrips. Asiatic onion leafminer can be controlled using 48% chlorpyrifos, 50% Diquling, 52.5% Nurelle, and other appropriate pesticides. Onion thrips can be controlled using 10% imidacloprid, 20% fenvalerate, 2.5% deltamethrin, 40% Dimethoate, or other suitable insecticides.

6. Harvesting

The bulbs, leaves, and garlic scape can all be used as food. Due to the different parts harvested and their usage, the timing, methods, and yield of harvesting also vary.

(1) Green garlic is consumed for its tender leaves and pseudostems. In the Yangtze River basin, after sowing from July to August, they can be harvested from October of the same year to the following spring. However, as summer approaches, the leaf tissues gradually age, and the fiber content increases, making them unsuitable for consumption as green garlic. The method of harvesting is mostly pulling them up once with the roots. They can also be harvested once in winter when the plants are about 30 cm tall, cutting them about 3–5 cm above the ground, at the base of the pseudostem. After harvesting, with enhanced fertilizer and water management, new leaves can regrow, they can be harvested once again in the period of February to March of the following year. When cultivating green garlic, the planting is very dense, and the yield of green garlic is also higher. For a single harvest, the yield per hectare is 30,000–40,000 kg, and with two harvests, the total yields can reach 50,000–60,000 kg.

(2) The harvesting of garlic scapes. Harvesting garlic scapes is closely related to harvesting garlic bulbs. Garlic scapes and the sprouting of garlic cloves

occur almost simultaneously. While garlic bulbs are harvested after they mature, garlic scapes elongate rapidly after sprouting and should be harvested promptly; otherwise, it can affect the yield of the garlic bulbs. The yields can reach 2,500–3,000 kg per hm^2.

(3) The harvesting of garlic bulbs. Garlic bulbs can be harvested 20–30 days after harvesting the garlic scapes. If garlic scapes are not harvested, garlic bulbs will still grow, but the yield may decrease by more than 15%. The harvesting season should coincide with a period of higher rainfall. If the harvest is delayed, the garlic bulbs can easily rot, and once harvested, they may also separate and have reduced storage life.

After harvesting, the garlic should be left to air-dry in the field for several days. Then, the bulbs should be bundled together and stored in a cool, shaded place. Like onions, garlic bulbs also have a certain dormancy period, and this dormancy period is longer than that of onions.

V. Common Cultivation Issues and Their Prevention Strategies

1. Horsetail garlic

This refers to the phenomenon in which, as the garlic scape emerges, a circle of small garlic leaves (5–7 small leaves) grows around the garlic scape, giving the whole garlic plant the appearance of a "horse's tail". Primary reasons: This occurs due to planting too early, insufficient planting depth, inadequate soil covering, excessive nitrogen fertilizer application, unsuitable varieties, and the occurrence of pests and diseases, among other factors.

Prevention and control methods: Select the appropriate planting time, choose suitable varieties, and improve water and fertilizer management. Especially control the amount of nitrogen fertilizer applied, pay attention to planting depth and soil coverage thickness, and promptly prevent garlic leaf blight and purple spot disease.

2. Single clove garlic

The bulb doesn't separate into garlic cloves, and this is mainly due to planting secondary bulbs, using very small seed cloves, inappropriate sowing time, insufficient soil fertility, planting too deep, planting too densely, and improper management of fertilization and watering.

Prevention and control methods: use suitable garlic seed, ensure the right size of seed cloves, provide sufficient basal fertilizer, focus on high-quality sowing, and enhance water and fertilizer management.

3. Scattered clove garlic

The garlic cloves formed by the bulbs fall off before they are harvested. The main reasons are early sowing, improper water and fertilizer management, and late harvesting.

Prevention and control methods: Choose the appropriate sowing period, select suitable varieties, improve fertilization and irrigation management, and harvest in a timely manner.

4. Double clove garlic

The lateral buds of the bulbs form garlic cloves, which sprout again to create a secondary level of cloves. The main reason is improper low-temperature conditions and handling before overwintering.

Prevention and control methods: Pay attention to the sowing period and control irrigation and fertilization.

「**Summary of Sub-context**」

This sub-context mainly focuses on the reproduction methods and field management of garlic, with an emphasis on selecting the right sowing period and giving importance to field management, particularly in terms of water and fertilization management.

「**Expanded Knowledge**」

The Benefits of Allicin

The benefit of allicin mainly manifests in three major aspects:

(1) Preventing and treating cardiovascular diseases such as hypertension, hyperlipidemia, hyperglycemia, coronary heart disease, dizziness, headaches, and post-stroke sequelae caused by atherosclerosis.

(2) Anti-inflammatory, antibacterial, antiviral, boosting the immune system, regulating the digestive system, preventing colds and respiratory infections.

(3) Anti-mutagenic, stimulating the production of interferon by immune-active cells, clearing or inhibiting carcinogens, suppressing malignant cell proliferation, and preventing tumors.

Sub-context 3 Production of Onion

Onion, also known as bulb onion or common onion, is a biennial herbaceous plant in the Amaryllidaceae family. It forms a bulb consisting of fleshy scales and a bulbil. The chromosome number is 2n=2x=16. It originated in the Central Asian region, with the Near East and the Mediterranean coast being the second center of origin. It was introduced to China around the beginning of the 20th century, and it is now cultivated in various regions. Onions are grown for their large fleshy bulbs, which are rich in protein, vitamins, and minerals, with a higher sugar content than scallions. They have a mild flavor and are known for their good taste. They can be consumed fresh or processed into dehydrated vegetables. Small-sized varieties are often used for pickling.

I. Biological Characteristics

1. Morphological characteristics in botany

(1) Roots: The onion plant has a fibrous and shallow root system with very few root hairs. The roots are mainly located at the base of the shortened stem disc, and their depth and lateral expansion range are only about 30–40 cm. The primary root mass is concentrated in the top 20 cm of the soil, forming a shallow root system in the soil. The root system has relatively weak drought resistance and limited absorption capacity.

(2) Stems: The true stem of an onion is a short, compressed, flattened, conical stem disc located at the base of the bulb. The upper part of this stem disc bears cylindrical leaf sheaths and buds, while the lower part supports fibrous roots. During the reproductive growth phase, the growth cone differentiates into floral buds, leading to the emergence of flower stalks.

(3) Leaves: The leaves are hollow, with a semicircular cross-section, consisting of leaf sheaths and tubular leaf blades. They grow upright, and the leaf blade is tubular with a concave belly. The leaf sheaths collectively form a pseudostem, which thickens at the base, and gradually swells to form flattened, round, or elongated bulbs. Inside the bulb, there are enlarged lateral buds, and the outer layers consist of dry, membranous scales, which can have colors like purple, yellow, green, or white. The weight of the bulb depends on the number of thickened leaf sheaths and their thickness, as well as the number of lateral buds (bulbils) inside the bulb. The outermost 1–3 layers of leaf sheaths at the base of the bulb transfer nutrients inward and shrink into membranous scales before the bulb matures. This helps protect the inner layers of scales and reduces transpiration, allowing onions to be stored for an extended period.

(4) Flowers: Onions typically produce flower stalks and bloom the following spring after bulb planting, then set seeds in the summer. After the onions bolt, each flower stalk has an umbel with a membranous involucre at the top. On top of this, floret grows, mostly pale purple or nearly white in color, and they undergo cross-pollination.

(5) Fruits and seeds: The fruit is a two-celled capsule that splits open when mature. The seeds are tiny, with a hard outer coat, often wrinkled. The seed coat is black and shield-shaped, with a thousand-seed weight of 3 to 4 g, and they have a short lifespan.

2. Requirements for environmental conditions

(1) Temperature: Onions have a strong temperature adaptability. Seeds and bulbs germinate slowly at 3–5°C, while germination accelerates at above 12°C. The optimal temperature for seedling growth is 12–20°C, and for leaf growth, it is 18–20°C. However, robust seedlings can tolerate temperatures as low as -6– -7°C.

The bulbing of the bulbs requires higher temperatures. Bulbs won't expand below 15°C. They start to expand at 15–21°C, and the optimal temperature during the bulbing phase is 20–26°C. If the temperature is too high or below 3°C, the bulbs will go into dormancy. After the bulbs mature, they have strong temperature adaptability and can resist cold and heat, allowing them to be stored during hot summers.

Onion is a vernalization-required vegetable plant. The plants must reach a certain size and accumulate a certain amount of nutrients to go through the vernalization phase. Most varieties can vernalize at temperatures range of 2–5°C for 60–70 days, typically when the stem diameter of seedlings is greater than 0.5 cm. However, there can be differences among varieties, with some northern varieties requiring 100–130 days for vernalization. After flower bud differentiation, the onion plant requires relatively higher temperatures to initiate bolting and flowering.

(2) Light: Onions are considered long-day vegetable crops. Under long-day conditions, leaf growth is suppressed, and the bases of the leaf sheaths and bulbil start accumulating nutrients, leading to bulb development. Extending the number of sunshine hours can accelerate bulb formation and maturation. The lengths of daylight hours required vary among onion varieties. Some varieties require 13.5–15 hours of long daylight, while others may only need 13.5–15 hours of long daylight. In northern China, there are many late-maturing onion varieties that require long daylight hours. Therefore, when inducing varieties from one region to another, it is essential to consider whether the introduced variety is suitable for the local daylight conditions. Onion bolting and flowering also require long daylight conditions.

(3) Water moisture: Onions have shallow root systems with relatively low water absorption capacity, so they are not well-suited for excessively dry conditions. They require soils with a certain level of fertility and good water retention capacity for cultivation. Onions require adequate moisture during the germination stage, peak growth stage of young seedlings, and bulb enlargement stage. Especially during the stage of peak growth and bulb enlargement, but it is advisable to control irrigation slightly before overwintering of young seedlings and within the 1–2 weeks leading to the maturity of bulbs to ensure that the

bulb tissues are plump, hastening maturity, and enhancing the quality and storage capability of the produce.

Onion leaves and bulbs are drought-resistant, which is why they require lower air humidity during their growth period. Excessive air humidity or rainfall during the flowering period affects flowering and fruiting. The bulbs are also strong drought-resistant and can retain moisture in the fleshy bulbs for a long time in arid conditions, supporting the vitality of the young shoots.

(4) Soil and nutrients. Onions thrive when grown in well-draining, fertile soil with good water retention capabilities, ideally in neutral loamy soil or sandy loam soil with a pH level between 6.0 and 6.5. Sandy, acidic, and saline-alkaline soils are not suitable for onion cultivation.

Cultivating onions requires nutrient-rich soil. During the seedling stage, nitrogen is the primary nutrient needed. As the bulbs swell during the later stages, it's advisable to increase the application of phosphorus and potassium fertilizers, as this helps enhance both yields and quality.

II. Types and Varieties

Onions can be categorized into long-day and short-day varieties based on their daylight requirements for bulb formation. Additionally, they can be classified based on bulb morphology into regular onions, tillering onions, and shallots. In production, regular onions are cultivated, and based on the bulb's external colors, regular onions are further divided into the following types.

1. Red onions

Also known as purple onions, these are mostly late to mid-season varieties. The plant has strong growth, with upright leaves and a deep green leaf color. The outer skin of the bulb is purplish-red or pink, the fleshy scales have a slight red color, and the bulb is round or slightly oval in shape. The texture is crisp, tender, and juicy. The spiciness is strong, but the flesh of the scales is not as dense and tender as yellow onions, with slightly lower quality. The natural dormancy period of the scales is relatively short. They sprout earlier, have higher moisture content, and don't store as well as yellow onions. Red onions are mostly cultivated in cool regions during the spring, with strong cold resistance and disease resistance. The main cultivated varieties are usually local varieties.

2. Yellow onions

They are mostly early to mid-season varieties with outer skin ranging from copper-yellow to light yellow, and the flesh is slightly yellow. The bulbs are flat-round or round, sometimes high-round. The plants have vigorous growth, with open leaves, and light green leaf color. The yield is slightly lower than that of red onions. The flesh of the scales is dense, tender, and fine-grained, with a sweet and spicy taste, making it of high quality and suitable for use as raw material for dehydrated vegetables. They are generally short-day varieties and are suitable for winter cultivation in the southern hot regions. Traditional main cultivated varieties include Taiyang, Shoutuo, and Nijiala introduced from the United States, as well as Hongye No. 3, Gangcong 841, Dabao, and Huang'guan introduced from Japan. Yellow onions are mainly for export.

3. White onions

They are mostly early-maturing varieties with white outer skin and a slight greenish tint on the lower part of the pseudostem. The flesh of the scales is also white. The bulbs are relatively small, often flattened and round. The flesh of the scales is tender and fine, and they have excellent quality. They are often used as raw materials for dehydrated vegetables or canned food ingredients. White onions have low yields, weak disease resistance, and are prone to premature bolting, making them rarely cultivated in production.

III. Cultivation Season and Methods

Cultivation of onions in southern regions is primarily for export and is concentrated in warm areas. Typically, they are sown in autumn and harvested in spring. However, due to market demand, some southern hot regions take advantage of the relatively high winter temperatures to cultivate onions with a summer planting and winter harvesting cycle. They are available on the market early, and their high prices lead to relatively high economic returns. The actual situation depends on the specific local practices and conditions.

IV. Cultivation Techniques

1. Plowing and fertilization

Onion root systems are distributed within the top 30 cm of the soil, and they have a weak ability to absorb nutrients and water. To promote the development of lateral roots and the absorption of nutrients, proper deep plowing is necessary. It is advisable to use sandy loam with high organic matter content and good water retention for onion cultivation. Onions do not thrive in acidic soil. A certain amount of calcium superphosphate can be incorporated into the base fertilizer to adjust the acidity. Additionally, phosphorus plays a vital role in root development and young seedling growth of onions. Excessive nitrogen fertilizer can lead to elongated seedlings, making it difficult for them to establish roots and grow well after transplanting.

2. Sowing and seedling raising

Onions are sown in autumn and overwinter as seedlings. If sown too early, during the overwintering period, the seedlings may grow too large, increasing the likelihood of premature bolting in the following year. However, if sown too late, although there won't be premature bolting in the following year, the plants will be too small when the bulbs start to grow, which affects the yield. The specific sowing period varies depending on the local climate conditions. The sowing period is later as they move south and earlier as they move north. In the Yangtze-Huai River basins, it is not advisable to sow before late September, while in Hangzhou and Shanghai, sowing is typically done in early October.

Due to the small size of onion seeds, the seedbed should be loose, fertile, and have good water-retention properties to facilitate the emergence of cotyledons. Generally, it is sown by dry seeding. Seeding on the seedbed should be even and sparse. Seeding about 60 seeds in a 100 cm^2 area can be used for a field 15 times larger. After sowing, cover with a layer of potting soil or wood ash, and then add a layer of straw or wheat straw on top.

3. Transplanting

The onion is usually transplanted in the mid- to late November, at the latest in early December. This way, the onions can have acclimated and grown before the harsh winter sets in, preventing frost damage.

During transplanting, it is important to select and grade the seedlings. Generally, seedlings aged 50–55 days, with stem thickness of 0.6–0.8 cm, and plant height of 25 cm, with "three leaves and one heart", are preferred. The quality and size of the seedlings are related to the possibility of premature bolting and, more importantly, to the growth of the plants and bulb yield after transplanting, so it's best to classify them. Onions have upright leaves, and densely planting them results in a significant increase in yield. In various regions, the typical row spacing is 15–20 cm, and the plant spacing is 10–15 cm. The planting depth is typically around 2–3 cm. Planting too deeply, with all the bulbs growing in the soil, can lead to deformities. Planting too shallowly, with too many bulbs exposed above the soil, may result in cracking, affecting the quality.

4. Field management

Field management of onions includes topdressing, irrigation, intertillage, and disease and pest control. Apart from the base fertilization, multiple rounds of topdressing are necessary. Timely and staged topdressing is one of the keys to achieving high onion yields. Approximately two weeks after transplanting, the first round of topdressing is recommended, which generally involves the application of a full amount of phosphorus fertilizer and an appropriate amount of nitrogen and potassium fertilizer to promote root system and above-ground growth. As the weather warms (usually in March), the onion plants start growing vigorously, and a heavier topdressing is required. From April to early May, which is the peak growing period for onions in the Yangtze River basin and the time of highest nutrient demand, it's advisable to apply nitrogen and potassium fertilizers once or twice more. Subsequently, as the bulbs mature, it's important to cease fertilization and reduce watering to prevent excessive moisture in the bulbs, as they become less storable. Additionally, considering the higher sulfur requirements of the plants, it's advisable to supplement with some sulfur fertilizer as needed.

Onions have varying water requirements at different stages of growth and development. In regions with abundant rainfall, especially in the southern parts, topdressing (manure) is generally combined with a reduced amount of watering. During the spring and plum rain seasons, proper drainage should be taken into account. Onion root systems are shallow, so intertillage should be kept shallow, generally not exceeding 3 cm, to avoid damaging the roots.

5. Pest and disease control

The primary diseases affecting onions are downy mildew and soft rot, while the main pest is the liriomyza. Timely prevention and control are essential. Downy mildew is the most severe disease for onions and can occur at various growth stages. It can be treated with sprays of Bordeaux mixture, Ridomil manganese-zinc, Dithane manganese-zinc, ethoprophos aluminum, chlorothalonil, etc. Below-ground pests, like white grubs (larvae of the seed fly), pose a threat to the roots. Root drenching with ammonia water and insecticides like dipterex can be somewhat effective.

6. Harvesting

In most regions along the Yangtze River, onion harvesting typically begins in the first half of June. Harvesting too early when the bulbs are not fully mature and have higher moisture content results in lower yields and reduced storage life. Harvesting too late can lead to rot, as onions tend to sprout early. The ideal time for harvesting is when the onion base has 2–3 leaves starting to turn yellow, and the pseudostem softens and starts lodging. At this stage, the majority of the leaves have partially withered but are not completely dried out. The harvesting method involves pulling the onions out by the roots on sunny days and letting them sun-dry in the field for 3–4 days to dry the outer skin, but without excessive exposure. It's recommended to stop watering a week before harvesting to improve bulb storage quality. To prevent sprouting during storage, a 0.25% solution of maleic hydrazide (MH) can be sprayed on the leaves before they have fully withered. Onions treated with MH are only suitable for consumption and should not be used for replanting. During harvesting, it's important to minimize bulb damage to reduce the risk of rot due to infections through wounds. If the onions are intended for fresh consumption, they can be cut about 6–10 cm above the pseudostem after sun-drying. The leaves are not cut for onions intended for hanging storage; instead, they are bundled and hung in a well-ventilated area.

Onions are vegetables with good storage potential, having a certain dormancy period (which varies by varieties). Under normal ventilation conditions, they can be stored for 5–6 months. After sprouting, they lose their edibility.

V. Common Cultivation Issues and Their Prevention Strategies

1. Premature bolting

The phenomenon of onions bolting before the bulbs reach maturity is called premature bolting. The primary reason for premature bolting is when young seedlings, with small bulb diameters exceeding 0.8 cm and 5–6 leaves, experience continuous low temperatures, leading to bolting under long daylight hours and high temperatures. This is mainly due to early autumn sowing, excessive irrigation, and fertilization during the seedling stage, causing excessive growth, and insufficient water and fertilizer in the anaphase, leading to early flower bud differentiation. Key prevention measures include selecting varieties less prone to bolting, sowing at the appropriate time, grading and transplanting young seedlings, nurturing robust seedlings, optimizing irrigation and fertilization management, applying a 0.2% ethephon solution during the seedling stage, and promptly removing flower scapes.

2. Improper variety induction

Incorrectly inducing onion varieties can result in low yields and poor quality. Primary reasons: Onion bulb formation is sensitive to temperature and daylight hours. Northern medium- and late-maturing varieties require over 1 hour of high-temperature light exposure for bulb enlargement, while southern medium- and early-maturing varieties need daylight conditions within 13 hours to form bulbs. When northern varieties are introduced to the south under shorter daylight conditions, bulbs fail to enlarge and only produce leaves. Conversely, southern varieties introduced to the north can exhibit early bulb enlargement, leading to premature maturity and small bulb size due to underdeveloped leaves, resulting in lower yields. To prevent this issue, it's advisable to introduce varieties from different latitudes, sow at the right time, and enhance overall management.

「**Summary of Sub-context**」

In this sub-context, the key focus is on the appropriate sowing period and the management of fertilization and irrigation before overwintering during onion cultivation. It is also important to strengthen fertilization and irrigation management in early spring. This approach not only helps prevent premature bolting but also increases yields. This approach serves a dual purpose: preventing premature bolting on one hand and increasing yields on the other.

「**Expanded Knowledge**」

The Benefits of Onions

Onions are a common and affordable staple in many households. While some people in China may be wary of their distinct pungent aroma, onions are hailed as the "queen of vegetables" in other parts of the world, boasting considerable nutritional value.

(1) Onions contain the micronutrient selenium, a potent antioxidant that can eliminate free radicals in the body, enhance cell vitality and metabolic capabilities, and have cancer-fighting and anti-aging properties.

(2) Onions also contain a good amount of calcium. Recent research by Swiss scientists suggests that regular onion consumption can increase bone density and aid in the prevention of osteoporosis.

(3) Onions are rich in natural antimicrobial substances like allicin, which grants them significant antibacterial properties. Chewing raw onions can help prevent colds.

(4) They are the only vegetable containing prostaglandin A. Prostaglandin A can dilate blood vessels, reduce blood viscosity, thereby lowering blood pressure, increasing coronary blood flow, and has a preventive effect on throm bosis. Regular consumption has a beneficial impact on individuals with hypertension, high blood lipids, and cardiovascular diseases.

「Reviewing and Thinking Questions」

I. Explanation of Terms

1. Softening cultivation.
2. Single clove garlic.
3. Double clove garlic.

II. Gap Filling

1. Onion is a ________ plant in the ________ genus of the ________ family. It forms a bulb consisting of fleshy scales and a bulbil.

2. The onion's flower scape is ________, with a swollen central portion, bearing an inflorescence, facilitating ________ -pollination.

3. Onion scales are composed of ________ and a________.

4. The optimum temperature for onion bulb formation is around ________ °C, and when the temperature exceeds________°C, bulb growth ceases, and the onion enters a state of physiological dormancy.

5. Most onion varieties require ________ days of exposure to temperatures around 2–5°C to complete vernalization. However, southern varieties need about ________ days at temperatures around 9–10°C, while northern varieties require approximately ________ days at temperatures around 3–5°C to achieve vernalization.

6. Onions thrive when grown in well-draining, fertile soil with good water retention capabilities, ideally in ________ soil or ________ soil with a pH level between ________ and ________.

7. Cultivating onions requires nutrient-rich soil. During the seedling stage, ________ is the primary nutrient needed. As the bulbs swell during the later stages, it's advisable to increase the application of ________ fertilizers, as this helps enhance both yields and quality.

8. The growth of onions, from seeding to seed maturity, comprises two main stages: ________ and ________. These stages encompass six periods: ________, ________, ________, ________, ________, and ________.

9. Generally, seedlings aged ________ days, with stem thickness of________ cm, and plant height of ________ cm.

10. The bulb enlargement period refers to the time frame from ________ to ________, during which the plant experiences significant growth, and the bulb gradually forms.

III. Choice Questions

1. According to the water requirements of the Allium family, it belongs to (　　).

A. Moisture-loving vegetables　　B. Semi-moisture-loving vegetables

C. Semi-drought-tolerant vegetables　D. Drought-tolerant vegetables

2. Onions can be categorized based on skin colors into (　　).

A. 3 types　　B. 4 types　　C. 5 types　　D. 6 types

3. The incorrect attention to detail during the harvest of yellow chives is (　　).

A. The knife used for harvesting should be sharp, and the cut should not be too deep

B. After cutting, it's advisable to immediately water the plants to promote growth

C. During the growth of yellow chives, harvesting can be done based on market demand, either by waiting or harvesting early

D. It's best to dry the harvested yellow chives in the sunlight for a period of time

4. When sowing annual chives in spring, they can begin to tiller (　　).

A. When the chive plants have grown 3–4 leaves

B. When the chive plants have grown 5–6 leaves

C. When the chive plants have grown 7–8 leaves

D. When the chive plants have grown 9–10 leaves

5. The vegetable belonging to the Liliaceae family is (　　).

A. Cabbage　　B. Chinese cabbage

C. Celery　　D. Garlic

6. The favorable temperature and light conditions for onion bulb development (　　).

A. Low temperature with long daylight

B. Low temperature with short daylight

C. High temperature with long daylight

D. High temperature with short daylight

IV. Thinking and Answering

1. What are the common biological characteristics and cultivation habits of Allium vegetables?

2. How should water and fertilizer management be carried out during garlic cultivation? What are the common issues in garlic cultivation? How can they be prevented?

3. How is Chinese chive propagated through sowing? How is Chinese chive etiolation cultivation performed?

4. What are the reasons for the premature bolting of onions, and how can it be avoided?

Learning Context 8　Production of Tuberous Vegetables

「**Learning Objectives in This Context**」

Master the main types and common characteristics of tuberous vegetables, the biological characteristics of the main types, and basic knowledge of tuberous vegetable production. Be able to choose the appropriate varieties of tuberous vegetables based on seasons. Be capable of creating production plans for tuberous vegetables and carrying out production correctly.

「**Analysis of Tasks in This Context**」

Tuberous vegetable cultivation involves soil preparation, basal fertilization, seed potato treatment, sowing, nutrient and water management, and pest and disease control.

「**Introduction**」

Tuberous vegetables, including common varieties like potatoes, yams, and sweet potatoes, play a significant role in people's life. Ginger is an essential spice, and tuberous vegetables are also important for export and generating revenue.

Overview

I. Types of Tuberous Vegetables

Tuberous vegetables include potatoes, yam, ginger, taro, *Helianthus tuberosus*, Chinese artichoke, and yam bean (Fig. 8-1). Their product organs are carbohydrate-rich underground tubers, rhizomes, bulbs, and corms, which have good storage and transport qualities, enabling a year-round balanced supply. These vegetables belong to different plant families, but they share many common characteristics in terms of biological traits and cultivation techniques.

Potato Yam Ginger

Fig. 8-1 Various Tuberous Vegetables

Tuberous vegetables primarily rely on asexual reproduction, which requires a significant quantity of planting materials, but their reproductive rate is low. After sowing, the seed pieces first sprout, and then develop adventitious roots from the sprout. As a result, the conditions for sprouting are stringent, and it takes a relatively long time. The product organs need to form in underground, dark conditions, and they require ample sunlight and significant day-night temperature variations. Therefore, when cultivating tuberous vegetables, it is essential to have loose and well-drained soil rich in organic matter. After sowing, the soil should retain moisture for an extended period, and the temperature should be suitable to facilitate root growth and emergence. During the seedling and sprouting stages, it is necessary to combine watering and fertilization, improve soil cultivation, and gradually mound the soil to create favorable conditions for the development of product organs.

II. Common Characteristics of the Cultivation of Tuberous Vegetables

(1) They all use asexual reproduction, require a large amount of planting materials, and have a low reproductive rate.

(2) The conditions for root development are stringent, and it takes a relatively long time.

(3) They require loose and well-drained soil rich in organic matter, and the mounding of soil creates dark conditions.

(4) During the peak period of product organ formation, strong light and significant day-night temperature variations are required.

Sub-context 1 Production of Potato

Potato is an annual herbaceous plant in the *Solanum* genus of the Solanaceae family. It is originally from the high mountain regions of South America. In China, potatoes are typically grown as both a staple crop and a vegetable in the high-altitude regions of the Northeast, Northwest, and Southwest. In the North China and the Yangtze-Huai River basins, they are primarily cultivated as vegetable. Potatoes are also used as a raw material for producing starch, glucose, alcohol, and other products. Potatoes are primarily grown for their underground tubers, which are rich in starch, sugar, protein, and various vitamins, making them a highly nutritious food source. Potatoes are suitable for intercropping with various other crops, including vegetables, grains, and fruit trees. This intercropping can effectively utilize land resources, enhance productivity, and is considered a space-efficient crop. Potatoes have good storage and transport qualities, making them an important crop for year-round vegetable supply and helping to fill gaps in availability.

I. Biological Characteristics

1. Botanical features

(1) Roots: Potatoes have a fibrous root system, and when the sprouts reach 3–4 cm in length, their base produces primary roots (bud eye roots), which form the main absorption root system. The primary roots typically grow horizontally for about 30 cm before extending vertically downward to a depth of 60–70 cm. Creeping stems are roots that grow around the underground stem nodes, mostly distributed in the surface layers of the soil.

(2) Stems: The potato plant's stem is divided into above-ground and below-ground parts. The aerial stem is mostly upright, with a rhomboid cross-section and green or sometimes purple spots. There are wavy or straight stem wings attached to the stem, which are distinguishing features for identifying varieties. The underground stem includes the lower part of the main stem, creeping stems, and tubers. The lower part of the main stem underground consists of 6–8 nodes. The leaves on these nodes degenerate into scale-like structures, and the creeping stem emerges from the leaf veins. At the tip of the stolon, 12–16 short, swollen internodes form tubers (Fig. 8-2). The connection point with the creeping stem is referred to as the potato tail or navel, and the other end is known as the potato top. The tubers are covered in many sprout eyes arranged in a spiral pattern. Sprout eyes on the potato tail are relatively sparse, while those on the potato terminal bud are denser, exhibiting stronger sprouting potential.

Fig. 8-2 Potato Tubers

(3) Leaves: The primary leaves are simple, heart-shaped, and single. Subsequent leaves are odd-pinnately compound, with stipules attached to the base, and their shape serves as a distinguishing characteristic for identifying varieties.

(4) Flowers, fruits, and seeds: Umbel or compound umbel with five-lobed, rotate corollas. The fruits are spherical or elliptical, with tiny, kidney-shaped seeds.

2. Requirements for environmental conditions

(1) Temperature: Potatoes are not cold-tolerant or heat-tolerant. They prefer cool and temperate climates. The optimal temperature for sprouting is 12–18°C. Stem elongation, creeping stem formation, and leaf expansion all require higher temperatures, with around 21°C being optimal. Tuber enlargement requires a certain diurnal temperature difference. The optimal day temperature is 17–19°C, the night temperature is 12–14°C, and the soil temperature is 16–18°C. When the temperature of the potato stem is below 2°C or above 29°C, tuber growth ceases. Excessively high night temperatures and soil temperatures can both inhibit tuber formation.

(2) Light: Potato sprouting requires darkness. Light can inhibit sprout elongation, causing them to thicken, promote tissue hardening, and stimulate pigment production. Long daylight hours during the seedling and sprouting stages are beneficial for stem and leaf growth as well as the formation of creeping stems. During the tuberization phase, a short photoperiod results in faster tuber formation.

(3) Water moisture: The sprouting phase requires adequate soil moisture. To ensure timely emergence, it is most suitable for the soil below the planted seed potatoes to be moist while the soil above them is dry. Moderate drought during the prophase of seedling development, followed by maintaining soil moisture in the anaphase, is beneficial for seedling growth. During the sprouting phase, it is beneficial to have adequate soil moisture in the prophase, with a relative humidity of 70%–80%, to promote sprouting. In the anaphase, water should be controlled appropriately to prevent excessive stem and leaf growth, which can affect the transfer of the growth center. During the tuberization phase, a significant amount of water is needed. It requires ample and balanced watering to keep the soil consistently moist, with a relative humidity of around 80%.

(4)Soil and nutrients: Potatoes thrive in deep, loose, well-draining, organic-rich sandy loam soils with a slightly acidic pH ranging from 5.6–6. The growth of tubers requires well-aerated soil to ensure sufficient oxygen supply. Potatoes are high-yield crops and require a significant amount of fertilizers. To produce 1,000 kg of fresh potatoes, requires the absorption of 5–6 kg of nitrogen, 1–3 kg of phosphorus, and 12–13 kg of potassium. Avoid using fertilizers containing chloride ions in production.

II. Types and Varieties

Potatoes are classified into early-, medium-, and late-maturing varieties based on the maturity of their tubers. Early-maturing varieties take 50–70 days from emergence to tuber maturity, medium-maturing varieties take 80–90 days, and late-maturing varieties take more than 100 days. Early-maturing varieties are primarily found in the North China Plain and the middle and lower reaches of the Yangtze River. These varieties have short plants, lower yields, moderate starch content, poor storage ability, and shallow eyes. Medium-maturing and late-maturing varieties are mainly distributed in the Northeast, Northwest, and mountainous areas of the Southwest. These varieties have tall plants, high yields, higher starch content, good storage ability, and deep eyes. For double-cropping and intercropping, early-maturing varieties should be chosen to bring products to market earlier. When aiming for high yields and storage, medium-maturing and late-maturing varieties are preferred.

Based on the strength of potato tuber dormancy and the duration of the dormancy period, potato varieties can be classified into those with no dormancy, moderate dormancy, and long dormancy. Varieties with no dormancy period break dormancy approximately one month after 20–25 days of harvest, moderate dormancy varieties break dormancy around 2 months after harvest, and long dormancy varieties can take 3 months or more to break dormancy.

III. Cultivation Season and Methods

The timing of potato cultivation should be based on two temperature indicators: First, ensuring that after emergence, the lowest temperature remains above 0°C to avoid frost damage; Second, focusing on the period with soil temperatures around 16°C during tuberization. In China, in the Northeast and on the Plateau in the Northwest, single-season cultivation is adopted, which involves spring planting and autumn harvesting. In the Central Plains region, characterized by hot and rainy summers, two-season cultivation, both in spring and autumn, is commonly practiced. For spring potatoes, the suitable planting time is when soil temperatures stabilize at 5–7°C. For autumn potatoes, the planting period is determined based on the local date of the first potato-killing frost. This date is used to calculate the critical emergence date, which is typically 50–70 days before that frost date. The exact planting date is then determined based on the number of days required for seed potatoes to emerge after planting.

IV. Cultivation Techniques

1. Spring potato cultivation

(1) Land preparation and fertilization. It's advisable to choose flat terrain with fertile, loamy soil that has good drainage and is not prone to frost. Avoid consecutive planting with other Solanaceae crops, such as eggplants and tomatoes. The growth of the potato root system and tubers requires higher levels of oxygen compared to other crops. Therefore, choosing soil that is rich in organic matter and loose is especially significant for increasing yields.

The general fertilization principles for potatoes are to apply sufficient base fertilizer, early topdressing, and increased potassium fertilizer. Base fertilizer should account for 60%–70% of the total fertilizer requirement. Generally, in conjunction with plowing the field, 1,000–2,000 kg of base fertilizer per mu should be applied, with well-rotted compost made from cow and sheep manure and mixed with weed straw being the most suitable. Half to one-third of this should be incorporated into the plow layer during land preparation, while the remaining portion can be concentrated in furrows during planting.

(2) Early sprouting and timely planting. Choose disease-free, healthy tubers as seeds. When cutting the tubers, if any are found to be diseased, they should

be decisively eliminated, and the cutting knife should be disinfected (boiled in hot water, roasted over fire, or disinfected with 75% alcohol). When cutting the tubers, ensure that each piece has an eye, and try to use the topmost eyes as much as possible. This means trying to have more cut pieces with eyes at or near the top. The cut pieces should be as uniform in size as possible, generally weighing around 25 g each, forming a triangular shape with a lot of flesh.

15–20 days before planting, if the seed potatoes have not broken dormancy, they can undergo sprouting treatment. There are several methods for this, but generally, you can increase temperature and humidity, or use a chemical treatment. The chemical commonly used is gibberellin. Cut the tubers and soak them in a gibberellin solution with a concentration of 0.5–1 mg/L for 10 min. If you are planting whole potatoes, you can soak them in a 10 mg/L gibberellin solution for 10 min. After soaking, arrange the seed potatoes closely on clean, moist ground with the eyes facing upward, and sparsely cover them with clean, moist, fine soil, just enough to hide the potatoes. Cover them with a row cover during the day, ensuring that the maximum temperature does not exceed 20°C. Insulate them with straw at night to maintain warmth until the sprouts are about 1.5 cm long for planting. Using mulching film cultivation has a warming effect, typically raising the temperature by 3–4°C compared to open field conditions, which can advance planting by 11–15 days. In the middle and lower reaches of the Yangtze River region, late January is suitable for planting. Double mulch cultivation allows for planting in mid- to late December.

(3) Proper planting density. In mulching film cultivation, soil hilling is generally not needed. High ridge double-row planting is used with an average row spacing of 50 cm and a plant spacing of 23 cm, resulting in 6,000 hills per mu. The seeding rate is 150 kg per mu. For double film cultivation, a closer planting is recommended with larger rows spaced at 50 cm and smaller rows at 35 cm. Two rows constitute one ridge, with a plant spacing of 20–23 cm. This results in 8,000 hills per mu.

(4) Field management. At the emergence stage, it's important to uncover the film and transplant the seedlings. Seal the edges with fine soil, and in the case of cold snaps or low temperatures, take measures to provide insulation and protect the plants from cold. In double film cultivation, the temperature should be maintained

at around 20°C, avoiding temperatures above 20°C. After the last frost, remove the small arches or row covers.

When the majority of the seedlings have emerged, it's essential to apply seedling-boosting fertilizer. 15–20 kg of urea should be applied per mu. About 15 days later, provide additional sprouting-boosting fertilizer. Pay attention to adding phosphorus and potassium fertilizers. In mulching film cultivation, all fertilizers are applied in the initial basal application, and there is generally no need for additional topdressing.

Water management is usually combined with topdressing. However, it's important to maintain soil humidity during the prophase and metaphase of the sprouting period and tuber formation. In the anaphase of tuber development, soil moisture should be reduced.

Intertillage and hilling are usually combined with topdressing 2–3 times, which can prevent "green head" and improve tuber quality. When the plants emerge, perform deep intertillage and shallow hilling, and during the sprouting period, perform shallow intertillage and add thicker hilling. Before the crop closure stage, make sure to hill up the soil. If there is excessive elongation during the sprouting period 1–6 mg/L chlormequat can be applied for foliar spray.

Additionally, based on specific conditions, thinning and removing flower buds may also be performed.

(5) Pest and disease control. The main pests are aphids, leafhoppers, and black cutworms, primarily affecting the underground parts and roots of seedlings. Chemical control can be carried out using pesticides like 50% phoxim emulsifiable concentrates, 1,000-fold dilution of 80% DDVP emulsifiable concentrate or 1,000-fold dilution of 80%. Dipterex soluble powder can be sprayed or drenched. The main diseases include viral diseases, late blight, scab, bacteria wilt, and ring rot. Integrated field management is the primary approach to control these pests and diseases. Potato viral diseases are particularly serious, causing symptoms like striping, leaf curling, and fern leaf. Additionally, these diseases often cause varying degrees of stunting in potato plants, which can significantly impact both yields and quality. They are caused by various viruses, such as Potato X virus and Potato Y virus, and are mainly transmitted through fluids by aphids. Control measures

include selecting disease-free seed potatoes, producing virus-free seedlings from stem tips, using disease-free tubers from true potato seedlings, aphid control, and using methods like 300-fold dilution of 5% Zhibingling solution and 400-500-fold dilution of 20% Virus A (moroxydine hydrochloride with copper acetate) wettable powder during the outbreak of the disease.

Early blight disease can affect potato plants during both the seedling and mature stages. It primarily damages the leaves and tubers. Symptoms on the leaves include dark brown circular spots with concentric rings, and when the disease is severe, a black mold layer may form on the spots, leading to leaf withering and drop. On tubers, the disease causes slightly depressed, dark brown circular lesions with a light brown, sponge-like dry rot beneath the skin. To manage early blight disease, disease-free tubers should be selected for planting, improve cultivation practices with adequate organic and phosphorus-potassium fertilization, and use fungicides such as 500-fold dilution of 64% oxadixyl wettable powder, 600-fold dilution of 75% chlorothalonil wettable powder, or the Bordeaux mixture at a ratio of 1 ∶ 1 ∶ 200 during the disease outbreak.

Late blight, commonly known as the plague, affects the leaves, stems, and tubers of potato plants. Symptoms appear before flowering, with affected leaves showing irregular yellowish-brown spots. In damp conditions, a white, mold-like substance forms around the lesions, and the back of the leaves may have a dense white mold, which is a characteristic feature of this disease. Stems or leaf petioles may develop brown streaks when infected. Infected tubers display large, concave, purplish-brown lesions, with the tissue underneath the skin also turning brown. These lesions may expand or rot the surrounding area. The disease is caused by the pathogenic oomycete and infected potatoes can serve as the primary source of infection for the next year. Prevention and control measures include selecting disease-resistant varieties, using disease-free seed potatoes, early planting, choosing well-draining and loose soil, and applying fungicides like 200-fold dilution of 40% aluminum phosphide wettable powder, 58% metalaxyl-mancozeb wettable powder, or 500-fold dilution of 64% oxadixyl wettable powder.

(6) Harvesting. Early potatoes can be harvested and supplied at any time based

on market demand. However, for storage and transportation, the tubers should be allowed to fully enlarge, and they are typically harvested when the potato plants start to wither and turn yellow, which indicates that most of the foliage has changed from green to yellow. This should be done on a dry, sunny day to prevent excessive moisture on the potatoes, but attention should be paid to avoiding direct exposure to intense sunlight during harvesting to prevent damage to the tubers.

2. Autumn potato cultivation

Cultivating autumn potatoes in regions with a two-season system is carried out during a season characterized by high temperatures and abundant rainfall in the prophase, followed by low temperatures and frost in the anaphase. The growing period is shorter compared to spring potatoes, and because autumn potatoes are planted with seed tubers that have not undergone dormancy, these unfavorable factors can lead to issues like tuber rot and seedling death, resulting in reduced or even total crop failure. Therefore, to achieve successful autumn potato cultivation, strict adherence to the following technical points is essential: early harvesting of spring potatoes and proper storage of seed tubers, rigorous selection and elimination of deteriorated and dead seedlings, effective sprouting and seedling protection, and enhanced field management.

(1) Choose appropriate potato varieties and sources of seed tubers. Select varieties with a short physiological dormancy period (80–90 days), such as Taishan No. 1 and Zhengshu No. 2. There are two main sources of seed tubers: spring potatoes planted in autumn, as well as seed potatoes produced in sunbeds and disease-free micro-tubers. It is preferable to use seed potatoes produced in sunny beds and disease-free micro-tubers. For example, Taiguo No. 1 has an emergence rate of 79.8% and a yield of 11,115 kg per hm^2 when used as spring potatoes planted in autumn, an emergence rate of 90.83% and a yield of 13,680 kg per hm^2 when used as seed potatoes produced in sunbeds, and an emergence rate of 97.26% and a yield of 21,060 kg per hm^2 when used as a disease-free micro-tubers. Spring potatoes for seeding should be harvested 15–20 days in advance, ensuring that they have essentially completed the dormancy period at the time of planting.

(2) Use small whole potatoes for sprouting and planting. Planting with small

whole tubers weighing 20–30 g helps reduce rot and improve emergence rates. To promote sprouting, you can use a sand bed or immerse the tubers in well water at a reduced bed temperature. Another method is to treat them with a 10 μl/L solution of gibberellin for 10 min, then rinse with clean water. After sprouting, place the tubers on damp sand indoors for 2–3 days to let them grow green and robust, enhancing the resistance of young sprouts to adversity.

(3) Plant at the right time. For autumn potatoes, the ideal planting time in the middle and lower reaches of the Yangtze River is at the end of August or the beginning of September. If you're using them as seed potatoes, it's better to delay planting until the early to mid-September. In higher-altitude regions like Yunnan and Guizhou, where summer temperatures are lower, potatoes can be planted as early as July to August.

(4) Densely plant and sow shallowly. In regions where the second crop is grown, autumn cultivation is mainly for providing seed potatoes for the following spring planting. Implementing a crop rotation system over 2–3 years can reduce the yields by 20%–30% and make the crop more susceptible to diseases and pests. Apply 45,000 kg of high-quality organic fertilizer per hm^2 as a base fertilizer, and apply 45–75 kg of three-element compound fertilizer in the furrows at planting.

Autumn potatoes should be densely planted, typically with row spacing of 50 cm and plant spacing of 20–25 cm, resulting in 67,500–82,500 plants per hm^2. When planting, dig shallow furrows, 2–5 cm deep, with a width of 10 cm, and then irrigate to lower the soil temperature. Once the water has percolated, place the seed potatoes and cover them with the soil to form small ridges. In southern regions, you can dig furrows at the surface, place the seed potatoes in the furrow, gently cover them with soil, and mulch with straw to retain moisture and lower the temperature, which helps with seedling emergence.

(5) Strengthen field management. Autumn cultivation is done within areas not affected by frost. Late harvesting is beneficial for increasing yields and improving the percentage of large potatoes. In the southern regions, harvesting typically occurs from late November to early December. When using the harvested potatoes as seed potatoes, precautions must be taken to prevent freezing, and they should be properly

stored during the storage process.

V. Common Cultivation Issues and Their Prevention Strategies

1. Causes of potato seed degeneration

The degeneration of potato seeds occurs due to the continuous cultivation of potatoes, resulting in stunted plants, withered leaves, smaller tubers, and decreased yields. This phenomenon is known as seed degeneration in potatoes. Research from both domestic and international sources indicates that potato seed degeneration is caused by a combination of factors. The primary reasons include virus infections in plants and tubers, the influence of high temperatures during tuber formation, and the prolonged asexual reproduction of potatoes, leading to seed degeneration.

2. Key approaches to addressing seed degeneration

(1) Summer planting for seed preservation. In areas where potatoes are grown in a single season, storing seed potatoes and planting them during the summer can help mitigate the effects of seed degeneration. This strategy ensures that the tuber development phase occurs during the cooler autumn season, avoiding the detrimental impact of high summer temperatures. Summer planting is particularly effective for preserving early-maturing potato varieties. The suitable time for summer planting is from late June to early July.

(2) Sunbed propagation and autumn seed preservation. This method is commonly used in areas with two growing seasons. It involves cultivating original seed stock in sunbeds or polytunnels during the winter and spring seasons, which will be used for production in the following year. Sunbed planting takes place from November to February of the next year, with high-density planting (40 plants/m^2) used to propagate small seed tubers. Harvesting occurs from March to early May, followed by storage in a well-ventilated area at 25°C to encourage early sprouting or the production of more sprouts. These seed stocks are then used for production in autumn field planting.

(3) Three-season crop rotation for seed preservation. In certain regions like Qujing in Yunnan and Liulongsan in Guizhou, farmers take advantage of the seasonal differences in mountainous areas due to topography. They practice crop

cultivation in three distinct seasons: winter-spring (from mid-November to late May), spring-summer (from March to July), and autumn-winter (from August to early November). Seed potatoes are sourced from different locations in different seasons, leading to a three-season crop rotation for seed preservation.

(4) Virus-free seed potato propagation. Utilizing the method of shoot-tip culture to produce virus-free plantlets, propagating virus-free potatoes, and using them as the original seed ensures higher quality and reliability in seed potato production.

Propagation of virus-free or micro-tuber original seed provided by research institutions involves the following steps:

① Spring propagation of microtubers. From late February to early March, plant virus-free shoots in a sunlight greenhouse or a polytunnel. The soil should not have had any Solanaceae crops planted in the last 3–5 years, and it should be fertile and well-prepared. Maintain a greenhouse temperature of 6–20°C, with proper ventilation and cooling when it reaches 20°C. Harvest in early May, and each plant yields around 1.6–2.2 micro-tubers.

② Autumn propagation of micro-tuber original seed. In early August, sow the micro-tubers propagated during spring or purchase virus-free original seed potatoes. Soak them in wet sand 20 days before planting, and when the sprouts reach 5–7 cm, remove them for drying. Plant when the sprouts turn green, using row spacing of 50 cm and plant spacing of 20 cm. Harvest the original seed around the end of October, yielding approximately 1,500 kg per mu, which can be used for seed potatoes in an area of 10 mu in the following year.

③ Spring cultivation of virus-free potatoes. In spring, cultivate virus-free potatoes to obtain high-quality commercial potatoes while also selecting high-quality potato seeds for the fall. The cultivation techniques are similar to those used in regular production fields. However, the potatoes intended for seed must be harvested 20–25 days in advance, selecting small potatoes weighing 15–20 g. These selected small potatoes should be kept in a cool place for 4–5 days. Once the potato skins turn green, they are placed in ventilated storage for the summer.

Repeat this process, starting from obtaining virus-free micro-tuber original seed, and continue through field propagation, with a maximum of 3 years and 6 crops in the same field. If used continuously, the yields will gradually decrease each year.

「**Summary of Sub-context**」

This sub-context primarily focuses on using the vegetative organs of potatoes as planting materials, with an emphasis on the extended germination period and the relatively large quantity of seeds required. In spring, before sowing, it is essential to treat the seeds with processes such as germination induction and cutting. When planting in the autumn, measures to break dormancy are often necessary. Suitable cultivation methods include high ridges or high mounds, and it is crucial to implement effective field management practices.

「**Expanded Knowledge**」

Why can't Sprouted or Green Potatoes be Consumed?

Potatoes that have sprouted can produce a toxin called solanine (also known as glycoalkaloid). In good quality potatoes, there are only about 10 mg of solanine per 100 g. However, in sprouted, green, or rotten potatoes, the solanine content can increase by 50 times or more. Consuming a very small amount of solanine may not necessarily harm the human body. Still, if you ingest 200 mg of solanine in one go (approximately half an ounce of sprouted or green potatoes), symptoms can appear within 15 min to 3 hours. The earliest symptoms include itching in the mouth and throat, upper abdominal pain, as well as nausea, vomiting, diarrhea, and other symptoms. Those with milder symptoms may recover within 1 to 2 hours through the body's detoxification processes. If you ingest 300 to 400 mg or more of solanine, symptoms can be severe and include elevated body temperature, repeated vomiting leading to dehydration, dilated pupils, light sensitivity, ringing in the ears, seizures, difficult breathing, and low blood pressure. In very rare cases, individuals may experience respiratory failure and even die. Therefore, it is crucial to seek medical treatment promptly for individuals with severe symptoms.

Sub-context 2 Production of Ginger

Ginger is a perennial plant in the *Zingiber* genus of the Zingiberaceae family capable of forming underground fleshy rhizomes. In China, it is primarily grown as an annual crop. Ginger is originally from China and Southeast Asia, and it has a long history of cultivation in China. Nowadays, ginger is cultivated in almost all provinces of China, with Shandong, Zhejiang, and Guangdong being the major production areas. Ginger is consumed for its underground rhizomes, which contain substances such as gingerol, gingerone, zingiberene, and gingerol, giving it a unique spicy flavor. Ginger is used as a fragrant spice and can also be processed into dried ginger, ginger powder, ginger juice, ginger sugar, and ginger wine. It can be candied or pickled and is believed to have properties that warm the stomach, dispel cold, and induce sweating.

I. Biological Characteristics

1. Botanical features

(1) Roots: Ginger is a shallow-rooted plant, and its root system primarily extends within the top 33 cm of the soil. Ginger roots can be divided into two types: fibrous (or adventitious) roots and fleshy roots. The fibrous roots, originating from the base of the young shoot, are the primary absorption root system. The fleshy roots, which develop at the nodes of the ginger rhizome, serve both absorption and support functions.

(2) Stems: Ginger has both aerial and underground stems. The aerial stem grows upright, and when the shoot breaks through the soil, the growing point at the stem's tip is surrounded by leaf sheaths, forming a pseudostem. This pseudostem has hairs, a slightly purple base, and tender leaves. It reaches a height of about 80–100 cm. The underground stem, also known as the rhizome, is the organ where ginger stores nutrients. It has short and dense internodes. The ginger rhizormes have fleshy roots, fibrous roots, buds, and aerial stems and are the product-bearing organs. The skin of the underground rhizome is yellow or

pale yellow, and the flesh is also yellow or pale yellow. The ginger plant, after sprouting and producing shoots, forms an enlarged base at the shoot's bottom, known as the "mother ginger". The mother ginger is fleshy, and on it, new rhizomes develop.

(3) Leaves: The leaves of ginger are lance-shaped with parallel veins and possess a waxy texture. They have distinct tubular, leathery leaf sheaths. The leaves are green and arranged alternately in two rows.

(4) Flowers: Ginger plants produce yellow flowers in tropical regions. The flowers resemble those of the plant known as "Canna" and have irregular perianth. Ginger flowers are borne in spike-like inflorescences, but it is quite rare for them to develop into fruits.

2. Requirements for environmental conditions

(1) Temperature: Ginger is sensitive to temperature and prefers a warm, moist, and frost-free environment.

Ginger does not sprout when the daily average temperature is below 10°C. At 15°C, it can break dormancy in 15 days. Sprouting starts slowly at 16–20°C, but it's optimal at 22–25°C, taking about 25 days. At 30°C, it only takes 10 days, but the sprouts might be thin and weak. Treatment for sprouting is best done with varying temperatures: 20–23°C in the prophase, 25–28°C in the metaphase, and 20–22°C in the anaphase.

Ginger's optimal temperature for stem and leaf growth is 25–28°C. However, for root and rhizome growth, it's better to have lower night temperatures, with daytime temperatures at 22–25°C and nighttime temperatures at 17–18°C. Growth of stems and leaves is inhibited at temperatures above 35°C, leading to gradual plant death. Growth will slow down below 20°C and stop at 15°C. Frosts can cause stem and leaf damage.

(2) Light: Ginger is a shade-tolerant plant, but its light intensity requirements vary at different developmental stages. During the germination stage, darkness is required. In the seedling stage, moderate light intensity is needed. As the plant starts sprouting and enters the vigorous growth phase, it requires higher light intensity. For ginger, the light compensation point during the seedling stage is

800 lx, and the light saturation point is 25 klx. When light intensity exceeds 30 klx, the photosynthetic intensity of the plant decreases. Therefore, during the seedling stage, it's advisable to provide diffused and partially shaded light. When the plants are in the vigorous growth phase, they need around 80 klx of light. At this stage, it's essential to remove shading materials promptly to ensure efficient photosynthesis.

(3) Water moisture: Ginger is a shallow-rooted plant with underdeveloped root system, which means it cannot access deep soil moisture. It has a weak resistance to drought and has strict requirements for soil moisture. It requires consistently moist soil, with soil humidity levels ideally between 70% and 80% for optimal growth. During the germination period, it's essential for the soil to be adequately moist to promote root development and seedling emergence. In the seedling stage, the soil should be kept consistently moist. As the plant enters the vigorous growth phase, it demands a significant amount of water to support its rapid growth. However, excessively wet soil or excessive rainfall can lead to excessive elongation of the plants and make them more vulnerable to diseases.

(4) Soil and nutrients. Ginger has relatively flexible soil texture requirements. It can grow in sandy or loamy soils, but it thrives in deep, loose, fertile soils with good organic matter content, good aeration, and drainage. It is not suitable for cultivation in saline, alkaline, or waterlogged areas. Ginger is sensitive to soil pH value, and it grows best in neutral to slightly acidic soil with a pH value ranging from 5 to 7. The pH values that are too high or too low can negatively affect the growth of its stems and leaves.

Ginger is a plant that appreciates ample and well-balanced nutritions. It primarily absorbs potassium the most, followed by nitrogen, and phosphorus the least. It is sensitive to nitrogen and potassium. Nitrogen deficiency can result in stunted plants, fewer branches, and thin, pale leaves. Potassium deficiency can lead to premature aging of the plant and have a more significant impact on yields and quality. Therefore, providing adequate potassium and nitrogen is essential for ginger cultivation.

II. Types and Varieties

In China, there are many local varieties of ginger, and they can be classified into two types based on plant morphology and growth habits.

1. Sparse seedling type

These plants are tall and robust, with thick stems, few branches, deep green leaves, sparse nodes on the rhizomes, and large ginger pieces, often arranged in a single layer. Examples include Laiwu Ginger and Guangdong Shulun Ginger.

2. Dense seedling type

These plants have moderate growth, many branches, green leaves, numerous and dense nodes on the rhizomes, and ginger bulbs arranged in double or multiple layers. Examples include Laiwu Ginger, Guangdong Milun Fine Ginger, Zhejiang Red Claw Ginger, Zhejiang Yellow Claw Ginger, Anhui Tongling White Ginger, Guizhou Zunyi White Ginger, and Yunnan Yuxi Yellow Ginger.

① Red Claw Ginger: As a local variety in Zhejiang Province, it grows vigorously, reaching a height of 60 to 80 cm. The ginger bulbs are large, with pale yellow skin and waxy yellow flesh. The buds have a light red color, and they contain fewer fibers. The taste is spicy and the quality is excellent. Each individual plant's rhizome weighs 400–500 g.

② Yellow Claw Ginger: It is also a variety in Zhejiang Province. It is relatively short, and its rhizome joints have a pale yellow color. The buds do not have a red color, and the internodes are short and compact. The flesh is dense, and the taste is spicy, with excellent quality. Each individual plant's rhizome weighs around 250 g.

III. Cultivation Season and Methods

The appropriate cultivation season for ginger should meet the following conditions: The soil temperature at a depth of 5 cm should stay consistently above 15°C throughout the period from emergence to harvest, ensuring a growing period of at least 14 days and accumulating more than 1,200°C of effective accumulated temperature during this period. In production, it's important to aim for the period of rhizome formation to coincide with times when there are significant day-night temperature differences and suitable climatic conditions. With modern cultivation practices, it's possible to plan for earlier planting or delayed harvesting, but it's crucial to ensure that the micro-environment conditions are conducive to ginger growth. This means carefully managing factors like temperature, humidity, and light to optimize ginger production. In regions with frost-free, warm climates like Guangdong, Guangxi, and Yunnan, ginger can be planted from January to April without any cover. In the provinces along the Yangtze River, ginger is typically sown from Guyu to Lixia, which usually falls from late April to mid-May. If mulching film coverings are used, it can be done about 10 days earlier. In the cold regions of Northeast and Northwest China, ginger can be grown in sunlight greenhouses and plastic tunnels, resulting in high yields.

IV. Cultivation Techniques

1. Cultivation season

Ginger has a long growing season, and in many regions of China, it is commonly planted in the spring and harvested before the first frost. In general, in areas like Guangdong and Guangxi, ginger can be sown at any time between January and April. In the Jiangsu, Zhejiang, Anhui, and Hubei regions, sowing typically occurs from late April to early May, while in North China, it's usually in early to mid-May. If planting is done too early and the soil temperature is low, it can lead to delayed emergence and the risk of rotting and seedling death. If planting is done too late, it shortens the growing period and can result in reduced yields.

Using protected cultivation methods for ginger allows for early planting and delayed harvesting, ultimately increasing yields. For example, with mulching film cover, planting can be done about 25 days earlier than conventional planting. With greenhouse cover, planting can be advanced by around 50 days. Adding a greenhouse cover in the anaphase can also delay harvesting by about 15 days. In ginger greenhouse cultivation, early planting usually involves inducing sprouting in late November to early December, followed by planting (transplanting) in mid- to late January of the following year, with harvesting taking place from late May to early June.

2. Greenhouse cultivation techniques of ginger

(1) Sowing and seedling raising.

① Selecting ginger: Choose ginger rhizomes for planting that are pure in variety, matured from the previous year, with large and thick flesh, bright yellow skin, no shriveling, free from frost damage, diseases, pests, or mechanical injuries, and with multiple fresh and vigorous sprouts.

② Sun-drying ginger: Before inducing sprouting, place the selected ginger rhizomes in the sun on a clear day. Stop sun-drying when the ginger flesh becomes dry, turns white, and shows slight wrinkles. Sun-drying helps reduce the moisture content of the ginger and raises the temperature, promoting uniform sprouting. After sun-drying for 1–2 days, place the ginger indoors for 2–3 days, covering it with a straw mat to facilitate nutrient decomposition; this process is known as "enclosing ginger". Typically, after 2–3 cycles of sun-drying and enclosing, you can begin the sprouting process.

③ Sprouting ginger: Inducing early and uniform sprouting in ginger is a key technique for successful early cultivation in ginger greenhouses. There are various methods for sprouting ginger, and for early cultivation in ginger greenhouses, a hotbed in the greenhouse can be used for sprouting. Here are the specific steps:

a. Preparation of the sprouting bed. Dig a trench in the middle of the greenhouse, with a width of 1.2–1.5 m and a depth of 30–35 cm. The length of the bed depends on the quantity of ginger seeds. Spread a layer of fresh cow or pig manure as a heating material at the bottom of the bed, adding a small amount of water and compacting it to a thickness of approximately 15 cm. Cover the heating material with a layer of fine soil, about 4–5 cm thick.

b. Sprouting and management. Place the ginger seeds, which have undergone sun-drying, on the prepared sprouting bed. The thickness of the ginger seed pile should be 10–15 cm. Cover it with a layer of straw and then with plastic film, creating a row cover for insulation. Throughout the sprouting process, do not lift the plastic film. During the night, add a straw mat for insulation on top of the row cover.

c. Criteria for ending sprouting. During the sprouting process, ginger goes through stages such as germination, skin cracking, and the gradual appearance of rings. Generally, ginger is best for planting when the ginger shoots have reached the second or third ring stage. At this point, sprouting should be terminated, and the ginger can be transplanted. The sprouting period takes about 45 days.

(2) Land preparation and base fertilization.

Ginger has a weak root system and is sensitive to both drought and waterlogging, making it unsuitable for continuous cropping. For greenhouse early cultivation of ginger, it's advisable to choose a site with slightly elevated terrain, convenient irrigation and drainage, and deep and fertile sandy loam soil with good aeration. If possible, it's best to deep plow the soil before winter, allowing it to be exposed to sunlight and weathered. Approximately 30 days before transplanting, you should set up greenhouse coverings and plastic films, and deep plow the soil once more to raise soil temperature and reduce soil moisture content. Additionally, proper drainage between the greenhouses must be ensured. Ginger plants have high yield potential and require substantial nutrients. Adequate base fertilizer should be applied. In greenhouses, 1,700 kg of compost should be applied as the base fertilizer. Deeply mix the compost into the soil, level the beds, and create flat beds that are 1.5 m wide within 6 m-wide greenhouses.

(3) Transplanting.

In greenhouse cultivation of ginger, the primary goal is to harvest fresh ginger for the market. The growth period is relatively short, and the yield per plant is low, so it's important to plant ginger densely. Typically, you can plant 4 rows per bed, with a spacing of 25–30 cm between plants and approximately 30 cm between rows. In a standard greenhouse, you can plant around 2,800 ginger plants. After the sprouting of ginger is complete, you should proceed with transplanting. Before transplanting, water the transplant holes and apply an appropriate amount of wood ash. During transplanting, each ginger seed should weigh 50–75 g, and have 1–2 short and robust sprouts. Place the ginger seed flat in the transplant hole, with the ginger shoot slightly tilted downward to facilitate the development of new roots at the lower end. After transplanting, cover the ginger with 4–5 cm of soil, level the beds, and cover with mulching film while creating a row cover for insulation.

(4) Field management.

① Temperature management: Before ginger seedlings emerge, it's important to cover them with a straw mat on the row cover at night. After the ginger seedlings emerge in mid- to late February, you should puncture the mulching film promptly to allow the ginger shoots (seedlings) to extend out of the film to prevent scorching of the tender stems and leaves. For the row cover, you can practice uncovering during the day and covering at night. In the case of large greenhouses, the ventilation management depends on the weather conditions. The greenhouse temperature should not exceed 35°C, especially when soil humidity is relatively high. Proper ventilation is crucial to preventing excessive elongation. By late March, as temperatures rise, you can remove the row cover. In early May, remove the side film of the large greenhouse, while keeping the top film for coverage.

② Moisture management: After the ginger seedlings have emerged, it's essential to keep the soil moderately dry to encourage seedling growth and root system development. Throughout the entire growth period from emergence to harvest, maintain soil moisture, but be cautious to avoid excessive wetness and waterlogging.

③ Topdressing: Ginger has high nutrient requirements, so in addition to applying sufficient base fertilizer, it's important to provide timely topdressing. Topdressing is typically done in two stages: The first application should take place after the ginger seedlings have emerged and the main stems have fully extended. Apply 10 kg of imported compound fertilizer per mu. The second application occurs when the plants enter the tillering stage. Apply 10 kg of imported compound fertilizer and 5 kg of urea.

④ Shade management: Ginger prefers shade and is not tolerant of intense direct sunlight. Diffused light is more favorable for its growth. Towards the end of April, it's advisable to set up shade structures in the greenhouse. Shade structures are typically constructed using bamboo poles, wires, or other materials to create a flat canopy that's approximately 1 m high. Shade materials such as shade nets and straw mats are used for shading. On overcast and rainy days, shading is not necessary. On sunny days, shading should be applied from 9:00–16:00. This shade management helps create optimal growing conditions for ginger. You can also consider intercropping with crops like corn, loofah, or bitter gourd to make the most of the available space.

⑤ Weeding and soil cultivation: When ginger begins to tiller, remove the mulching film covering the soil and carry out the first weeding and soil cultivation. Subsequently, weeding and soil cultivation should be performed multiple times, combined, with fertilization and irrigation management. This helps maintain a clean and fertile growing environment for the ginger plants.

⑥ Pest and disease control: The main diseases affecting ginger are rot disease (commonly known as ginger wilt disease) and leaf blight. Ginger wilt disease is caused by bacterial pathogens and can be spread in the field through rainwater, runoff, and underground pests. It is more likely to occur in summer with high temperatures and heavy rainfall, especially when the temperature is above 20°C. Rainy weather, heavy and poorly drained soils, continuous cultivation over several years, and poor irrigation conditions can promote the spread of this disease. Symptoms include initially appearing as yellowish-brown water-soaked spots on the rhizomes, gradually enlarging, tissue softening and rotting, and the release of foul-smelling white fluid, leaving only empty husks. The roots turn yellowish-brown and rot, and the stems become dark purple. To control ginger wilt disease, you can treat ginger seeds by soaking them for 30 min in a 78% Jiangyining wettable powder, followed by incubation for 6 hours, or soak the ginger seeds in a 40% formalin solution at a 100-fold dilution for 6 hours. During the seedling stage, irrigate the furrows with a 300-fold dilution of 78% Jiangyining wettable powder, or use a 1,000-fold dilution of agricultural streptomycin sulfate wettable powder or a 3,000-fold dilution of neomycin sulfate solution or a 90% phosethyl-al soluble powder solution at a 300-fold dilution. Repeat this treatment every 10–15 days for 2–3 times. As for pests, the common ones are corn borers and striped flea beetles. Timely prevention and control measures should be taken to protect ginger plants from pest damage.

(5) Harvesting.

The harvest of ginger can be divided into three main categories: seed ginger, young ginger, and fresh ginger. Seed ginger is typically harvested along with fresh ginger when the growth cycle is complete. It can also be harvested at the anaphase of seedling growth, but care must be taken not to damage the seedlings. Young ginger is harvested during the period of vigorous rhizome growth when the ginger is tender and fresh. This is suitable for processing various food products. Fresh ginger is generally harvested just before the arrival of the first frost. About 3–4 days before harvesting, water the ginger once. When harvesting, you can pull out the entire ginger plant, shake off the soil, leave about 2 cm of aerial stem, and then use your hands or a knife to remove the roots. Store the ginger in a cool, damp place without the need for drying. For ginger grown in greenhouses, it's advisable to start harvesting early for the market. You can begin harvesting when the tillering has reached the third or fourth stage. Typically, this harvesting period starts in late May and extends into early June.

V. Common Cultivation Issues and Their Prevention Strategies

Common issues like ginger rot and seedling death can be attributed to two main reasons: ginger wilt disease (ginger wilt) and improper fertilization practices, including excessive fertilizer application.

Prevention and control methods: first, timely control of ginger wilt disease, and second, proper and balanced fertilization. Typically, during soil preparation, for every mu, apply 3,000 kg of well-rotted pig or cow manure, 30 kg of calcium superphosphate, and 15 kg of potassium fertilizer as the base fertilizer. When the ginger seedlings reach a height of approximately 30 cm, apply one round of fertilization, using 750–1,000 kg of pig manure water per mu. Around the Summer Solstice, when the ginger plants have grown 3–4 shoots, apply 50–100 kg of well-rotted sesame cake mixed with 300–500 kg of lime, or use 1,000–1,500 kg of pig manure water per mu. After fertilization, cover with a layer of fine soil or organic material. Around the time of the hottest part of the summer, apply 1,000–1,500 kg of pig manure and 4–5 kg of urea per mu.

「**Summary of Sub-context**」

In this sub-context, the focus is on introducing the seed ginger treatment, field management, and harvesting of ginger.

「**Expanded Knowledge**」

Reasons to Eat Ginger in Early Summer

(1) Appetite stimulation and digestive health promotion.

(2) Cooling effect, reducing heat, and refreshing.

(3) Antimicrobial and detoxification properties.

(4) Motion sickness prevention and nausea reduction.

Sub-context 3 Production of Taro

Taro, belonging to the *Colocasia* genus of the Araceae family, is a perennial herbaceous plant. Its edible part is the underground corm, and it is typically grown as an annual vegetable crop. Taro is originally native to the southern regions of Asia, including southern China, India, and the Malay Peninsula, especially in tropical swampy areas. Wild taro, the original species, grows in swampy areas. Through long-term natural selection and human cultivation, various types of taro have been developed, including water taro, taro suitable for both wet and dry conditions, and dryland taro. Wild taro typically has underdeveloped corms and petiole, a strong astringent taste, and in some cases, can be toxic and unsuitable for consumption. It usually has creeping stems that form small corms at the tips, primarily used for reproduction. Through natural and artificial selection, taro varieties with thick petioles for leaf consumption and well-developed corms for culinary use have been developed.

Taro is distributed worldwide, but is most commonly cultivated in China, Japan, and various Pacific islands. In China, it is primarily grown in the Pearl River basin and Taiwan. The Yangtze River basin and North China regions also cultivate taro, although to a lesser extent. Due to advancements in cultivation techniques, taro cultivation has expanded northwards. Taro is available for consumption from August to September, making it an important crop during this period. It is also known for its storage durability, ensuring a long supply period. Additionally, taro is versatile as both a vegetable and a source of starch.

I. Biological Characteristics

1. Botanical features

(1) Roots: Taro plants have white, fleshy, fibrous roots that grow from the nodes at the base of the corm, which include the mother taro and the daughter taros. These roots have few root hairs, limited absorption capacity, and are not well-suited for dry conditions.

(2) Stems: Taro stems are short and compact, forming underground corms. These corms can take on various shapes, including round, ovate, elliptical, or oblong. The corms display distinct leaf-shaped rings on their surface, with brown, scale-like hairs at the nodes, representing remnants of leaf sheaths. Axillary buds can develop into new corms on the nodes of the corms. Some varieties produce creeping stems that swell at the tips, forming new corms. The structure of the corm primarily consists of parenchyma cells in the basic tissues, including the cortex and pith. The vascular bundles are dispersed, with large vessels that connect to leaf vessels, extending directly to the vicinity of stomata and water pores. This adaptation is a response to the plant's preference for wet environments. New plants can sprout at the base of the seed corm, forming the mother taro, which can give rise to daughter taros.

(3) Leaves: Taro leaves are arranged alternately on the stem. They are large, 25–90 cm in length and 20–60 cm in width. Most commonly, they are shield-shaped, although some may be oval or slightly arrowhead-shaped, tapering gradually towards the tip. The leaf surface is covered in dense papillae, which trap

air and create air cushions, allowing water droplets to form into round beads that don't wet the leaf surface. Petioles can be long, measuring between 40 cm and 180 cm, and they can be either upright or spreading. The lower part of the stalk swells to form a sheath that encircles the stem. In the middle part, there is a groove, and the stalk can exhibit various colors, including green, red, purple, or dark purple, often serving as the basis for varietal names. Both the leaf blades and the petioles have noticeable air cavities, with poorly developed xylem. The petioles are long and hollow, and the leaves are large, although they are not particularly resistant to wind and rain.

(4) Flowers: Taro plants produce an inflorescence known as a spadix. The actual flowers are encased within the spathe, which is a leaf-like structure. Flowering is rare in temperate regions. In certain varieties, such as the purple-tipped taro and red taro found in Fujian Province, some individual plants may produce flowers. The fruit that develops after flowering is typically a berry.

2. Requirements for environmental conditions

(1) Temperature: Taro plants thrive in warm and humid environments. They grow well in regions with an average daily temperature of 21–27°C and an annual precipitation of around 2,500 mm. Corms begin sprouting at temperatures around 13–15°C, and for optimal growth, they require temperatures above 20°C, with the range of 27–30°C being ideal. Different taro varieties have varying temperature requirements. Some types of taro, like Duozi taro, are more tolerant of lower temperatures, while Kui taro requires high temperatures and an extended growing season to allow the corms to fully develop. As a result, Kui taro is commonly cultivated in the Pearl River basin in China, while the Yangtze and Yellow River basins are suitable for growing Duozi taro and Duotou taro.

(2) Light: Taro is relatively shade-tolerant. Strong sunlight, especially when combined with high temperatures and dry conditions, can lead to scorching of the leaves. Taro benefits from shorter periods of sunlight, which promotes the development of underground corms.

(3) Water moisture: Taro plants exhibit characteristics typical of aquatic plants in their leaf, root, and petiole tissues. Even when grown as dryland taro, it

is advisable to select areas with sufficient moisture. Throughout the growth period, particularly during the peak growth stage, taro should not be subjected to water shortages. Drought conditions can hinder growth, prevent leaves from developing fully, and result in significant yields reduction.

(4) Soil and nutrition: Taro plants are fond of fertile soil and have high nutrient requirements. During the peak growth stage, they should not face nutrient deficiencies. Adequate supplies of nitrogen and potassium fertilizers are beneficial for increasing yields and improving quality.

3. Growth, development, and corm formation

Taro reproduces using corms, known as seed taros. Under suitable temperature and humidity conditions, the corms sprout and develop into the first true leaf, which typically takes about 30 days from the start of germination.

After germination, seed corms produce new plants. As the plants grow, the shortened stem at the base of the terminal bud gradually swells to form a corm, known as the mother taro. Seed corms shrink and may even rot as they expend their nutrients. With each extension of the shortened stem, one leaf emerges above the ground, engaging in photosynthesis and nutrient production. Initially, the leaves are relatively small, but during the peak growth stage from June to August, they reach their maximum size, exerting the most significant influence on yields. Typically, it is advisable to maintain a robust set of 7–14 leaves to enhance assimilation efficiency.

Each segment of the mother taro has one axillary bud. Robust axillary buds in the middle and lower sections of the mother taro develop into corms, referred to as daughter taros. Daughter taros, like mother taros, emerge from the soil and produce leaves. They develop roots on the stem segments. Under favorable growth seasons and environmental conditions, they can continue to form the next generation of taro.

In regions south of the Yangtze River, the peak period for corm formation generally occurs from July to September. During this period, daughter taros swell, and the production of granddaughter and great-granddaughter taros continues. Growth slows down after October, and nutrients are transported to the corms, leading to an increase in starch content. The quantity and proportion of daughter taros and their quality vary depending on the types and varieties of taro. For example, in the case of Binglang taro, a type of Kui taro, there are fewer daughter taros, and the proportion of mother taros is high, resulting in the best quality. Next in quality are the daughter taros, while granddaughter taros tend to be more sticky. In Baigeng taro, a type of daughter taros, there are more granddaughter taros, and the proportion is higher than mother taros, leading to the best quality, while the quality of mother taros is inferior.

II. Types and Varieties

Taro has two main variants: leaf-type taro and corm-type taro. Leaf-type taro produces petioles that have no astringent taste or have a mild astringent taste. Examples include Guangdong Red-Stem Taro, Zhejiang Ningbo Water Taro, and Sichuan Wulong Leafy Taro. Corm-type taro, is primarily grown for its enlarged corms. This type can be further categorized into three types based on the development of mother taros, daughter taros, and their growth habits: Kui taro, Duozi taro, and Duotou taro.

1. Kui taros

The mother taro in this type is large, weighing up to 1.5–2 kg, and has better quality than the daughter taro. It has a strong aroma and a powdery texture. It thrives in high-temperature regions and is commonly cultivated in the Pearl River basin and Taiwan. This type includes the following varieties.

(1) Creeping Stem Kui Taro: The mother taro is large and of good quality. Creeping stems grow from the mother taro, and the swollen daughter taros at the top are not edible but are used for reproduction, such as Sichuan Yibin Chuangen Taro.

(2) Long Kui Taro: The mother taro is cylindrical, and the daughter taros have either long or short stalks, and they are edible. Examples include Fujian Tube Taro (long stalk) and Fujian Bamboo Taro (short stalk).

(3) Coarse Kui Taro: The mother taro in this type is elliptical and lacks a distinct stalk. Examples include Taiwan Mian Taro, Nuomi Taro, and Zhejiang Fenghua Huo Taro.

2. Duozi taros

Duozi taros have many daughter taros that are sessile and easily separable with better yields and quality than the mother taros. They generally have a sticky texture. Examples include: Aquatic taros such as Yichang White Taro (belonging to the green-stalk variety group) and Yichang Honghe Taro (belonging to the purple-stalk variety group); Dryland taros such as Shanghai Baigeng Taro, Guangzhou Baiya Taro, Fujian Qinggeng Wuya Taro (belonging to the green-stalk variety group), Guangdong Hongya Taro, Fujian Honggeng Wuniang Taro, and Taiwan Sow Taro; Taros suitable for both wet and dry conditions such as Changsha Baihe Taro (belonging to the green-stalk variety group) and Changsha Wuhe Taro (belonging to the purple-stalk variety group).

3. Duotou taros

Corms in this type grow in clusters, and there is no distinct difference between mother, daughter, and granddaughter taros. They are closely packed, and their texture is somewhere between powdery and sticky. Typically, they are dryland taros. Examples include: Dongjiu Mian Taro, Jiangxi Xinyu Goutou Taro, Zhejiang Jinhua Qie Taro (belonging to the green-stalk variety group), Fujian Changjiao Jiutou Taro, Sichuan Lianhua Taro (belonging to the purple-stalk variety group).

III. Cultivation Season and Methods

Taro requires high temperatures and has a long growth period, which is why it's mostly cultivated in open fields. The cultivation seasons vary significantly in different regions due to differences in latitude and altitude. In the Pearl River basin, Guangxi and Guangdong, taro is sown from February to March. In Sichuan and southern Fujian regions sowing starts in early March, while in the Yangtze River basin, sowing begins in early April. In Coastal areas of Shandong, sowing starts in early April. Sowing should take place when the soil temperature stabilizes at 8–10°C at a depth of 10 cm. It's advisable to sow early, provided there's no risk of freezing, as this promotes early root development and helps increase yields. Early-maturing varieties are typically harvested from July to August, while late-maturing varieties are harvested before the frost arrives, with a growth period of 200 days or more.

IV. Cultivation Techniques

1. Land selection and base fertilization

Taro requires soil that is rich in organic matter, deep, fertile, and capable of retaining water, such as loamy soil or clay. Although it's suitable for dryland cultivation, it's still advisable to choose relatively moist areas. Taro should be rotated with other crops and not planted continuously, with a recommended crop rotation cycle of at least 3 years.

Taro has a long growth period and requires substantial amounts of nutrients. For dryland taro, it's common to apply 37,500–45,000 kg of well-rotted organic and miscellaneous fertilizers per hm^2. Additionally, 300–450 kg of nitrogen-phosphorus-potassium compound fertilizer should be added, with an emphasis on potassium, as it can increase the starch content and aroma of the corms.

2. Seed taro selection and seedling raising

(1) Seed taro selection. Choose robust daughter taros in the middle of mother taros from healthy plants in disease-free fields. The selected single-seed taro should weigh 50 g or more, have well-developed terminal buds and corms, and exhibit uniform shapes. Avoid using corms from varieties such as Baitou, Luqing, and Changbin, which typically have less suitable characteristics for seed production. Baitou taros are often granddaughter taros without a top bud and exhibiting no scale-like hairs at the top, Luqing taros refer to the taros that have developed leaves, and Changbin taros are characterized by having daughter taros attached to the base of the mother taros. These taro tissues are not fully developed, and if used as seed taro, the seedlings will be weak, which will adversely affect the yield. Due to the continuous cropping of mother taro, daughter taro, and granddaughter taro, it is challenging to separate them. Therefore, it is necessary to divide them into several seed taros. The taro requires a large amount of seeds, with a general seeding rate of 50–200 kg per mu.

(2) Mother taro used as seeds. Mother taro used as seeds can be divided into using mother taro pieces and cultivating small mother taro seeds.

① Using mother taro pieces: Using the entire mother taro for seeding significantly increases yields, but it requires addressing issues like rotting seeds.

② Cultivating small mother taro seeds: Binglang taros have low reproduction rates, and some small taro seeds have low yields. To improve utilization, daughter taro can be grown for one year to become small mother taro seeds.

(3) Inducing sprouting and seedling cultivation. Taro has a long growth period, and inducing sprouting and seedling cultivation can extend the growing season and increase yields. Typically, in early spring, about 20–30 days in advance, taro seeds are cultivated on a cold bed. The bed temperature should be maintained at 20–25°C with suitable humidity. The bed soil should not be too thick, as this restricts root system penetration, making it easier for transplanting. When taro seedlings have grown to 4–5 cm and there is no frost in the open field, it's time to plant them.

3. Planting

Taro is shade-tolerant and can be densely planted for significant yields increase. To facilitate hilling, a wide-row narrow-planting method is commonly used. For planting Duozi taro, the row spacing is about 80 cm, with a plant spacing of approximately 20 cm. This results in planting 4,000–5,000 plants per mu. Taro should be planted deeply to facilitate corm growth. The depth of soil cover should be around 10 cm above the top of the seed corm, with the terminal bud slightly exposed. Planting too shallow can affect root development.

4. Field management

(1) Seedling stage management. Before and after seedling emergence, perform multiple times of intertillage, weed control, and soil loosening to increase soil temperature and promote root and shoot development. If any gaps in the seedlings are noticed, promptly fill them with new seedlings. When using mulching film mulch for cultivation, manually uncover the film when young shoots emerge to prevent heat damage. Afterward, cover the soil and press down around the film edge to enhance the warming and moisture retention effects of the mulching film.

(2) Fertilizer and water management. Taro requires a high amount of nutrients. In addition to the base fertilizer, it's essential to apply topdressing at different growth stages to promote growth and corm development. The principle for

topdressing is to provide lighter fertilization during the seedling stage and heavier fertilization during corm formation. Taro prefers moisture and dislikes drought. In the prophase when the temperature is lower and growth is slower, maintaining soil moisture is sufficient, but waterlogging should be avoided to prevent root system growth issues. During the metaphase and anaphase when growth is vigorous and corms are forming, adequate irrigation is necessary. Irrigation should be done in the morning or evening during hot weather. Avoid midday irrigation, which can harm the roots.

(3) Soil hilling. Soil hilling is essential for preventing terminal bud sprouting, promoting the expansion of daughter and granddaughter taro, enhancing lateral root growth, and improving absorption and drought resistance. Typically, as the above-ground part of the taro grows rapidly in June and the mother taro expands quickly, and as the daughter and granddaughter taros begin to form, it's time to apply soil hilling. Daughter taros of Duozi taros are prone to sprouting above the soil. You can combine soil hilling with intercropping and weeding 2–3 times, ensuring an even application around the corm each time to help maintain the proper taro shape.

5. Pest and disease control

The main diseases affecting taro are taro rot and taro blight, which occur more severely in hot climates. Control methods include implementing 3–4 years of crop rotation, using disease-free corms for planting, reducing mechanical damage to plants, and spraying Bordeaux mixture, Duojunling or other fungicides at the onset of disease. The primary pest is the tobacco cutworm, whose larvae damage the leaves. Insecticides like dipterex or deltamethrin can be used to control them.

6. Harvesting

The yellowing and wilting of the leaves are signs of mature corms. At this stage, the taro has a high starch content, good quality, and high yields. However, for market supply, harvesting can be done earlier or later. Early-maturing varieties in the Yangtze River basin are often harvested in August, while late-maturing varieties are harvested in October. Typically, you can expect to harvest 1,500–

2,500 kg per mu. After harvesting, remove the withered leaves, leaving the corms attached to the plant. Allow them to air dry for 1–2 days, and then store them in a cellar or storage area.

V. Common Cultivation Issues and Their Prevention Strategies

A common issue is "green tip exposure", which means the corms that are exposed on the surface of the soil turn green due to external light exposure and wind. In cultivation, as long as corms are exposed on the soil surface, it's essential to hill or cover them with soil promptly. With 2–3 hilling or covering cycles, the problem of green corms can generally be prevented.

「**Summary of Sub-context**」

The main focus of this sub-context is to understand the types and varieties of taros, cultivation techniques, and particularly the prevention and control of pests and diseases affecting taros.

「**Expanded Knowledge**」

The Benefits of Taro

In traditional Chinese medicine, taro is considered to have a sweet, pungent, and mild nature, and it enters the spleen and stomach meridians. It is known for its effects such as benefiting the stomach, relaxing the intestines, promoting bowel movements, nourishing the middle, benefiting the liver and kidneys, and enriching the essence and marrow. It can be used as an adjuvant treatment for conditions such as constipation, thyroid enlargement, scrofula, mastitis, insect bites, bee stings, intestinal blockages due to worms, and acute arthritis. However, it should be noted that taro should not be applied or used on healthy skin, as it can cause dermatitis. If this happens, gently wash it with ginger juice to alleviate the condition.

「**Reviewing and Thinking Questions**」

I. Explanation of Terms

1. Seed degeneration.
2. Enclosing ginger.
3. Air potato.

II. Gap Filling

1. Tuberous vegetables primarily rely on ________ reproduction, which requires a quantity of planting materials, but their reproductive rate is________.

2. Adequate supplies of ________ fertilizers are beneficial for increasing yields and improving quality of tuberous vegetables.

3. Potatoes require a significant amount of fertilizers, prefer ________ soil, and dislike________.

4. Potatoes are classified into ________, ________, and ________ -maturing varieties based on the maturity of their tubers.

5. For autumn-sown potatoes, newly harvested seed potatoes from the same year should be used, and they need to undergo ________ soaking to break dormancy.

6. There are wavy or straight ________ attached to the potato stem, which are distinguishing features for identifying varieties.

7. Ginger is a plant that appreciates ample and well-balanced nutrition. It primarily absorbs the most, followed by ________, and________ the least. It is sensitive to nitrogen and potassium.

8. During the sprouting process, ginger goes through stages such as________, ________ , and the________.

III. Choice Questions

1. The edible part of taro is (　　).

A. Fleshy stem　　B. Bulb

C. Tuber　　D. Corm

2. Among the following vegetables, ( ) uses the tuber as the reproductive organ

A. Sweet potatoes B. Yam

C. Ginger D. Yam bean

IV. Thinking and Answering

1. Based on the biological characteristics of potatoes and considering the local natural and cultivation conditions, develop measures for high-yield potato cultivation.

2. How to prevent genetic degeneration of potato varieties?

3. What are the steps involved in ginger seed treatment? What are the key technical aspects of field management?

Practical Training Tasks

I. Task Plan Sheet

<table>
<tr><td colspan="2">Learning context 1</td><td colspan="4">Vegetable production</td></tr>
<tr><td colspan="2">Work task</td><td colspan="4">Horticultural vegetable cultivation</td></tr>
<tr><td>No.</td><td>Plan implementation steps</td><td colspan="2">Plan implementation time</td><td colspan="2">Material preparation for implementation steps</td></tr>
<tr><td></td><td></td><td colspan="2"></td><td colspan="2"></td></tr>
<tr><td></td><td></td><td colspan="2"></td><td colspan="2"></td></tr>
<tr><td></td><td></td><td colspan="2"></td><td colspan="2"></td></tr>
<tr><td></td><td></td><td colspan="2"></td><td colspan="2"></td></tr>
<tr><td></td><td></td><td colspan="2"></td><td colspan="2"></td></tr>
<tr><td></td><td></td><td colspan="2"></td><td colspan="2"></td></tr>
<tr><td></td><td></td><td colspan="2"></td><td colspan="2"></td></tr>
<tr><td></td><td></td><td colspan="2"></td><td colspan="2"></td></tr>
<tr><td></td><td></td><td colspan="2"></td><td colspan="2"></td></tr>
<tr><td></td><td></td><td colspan="2"></td><td colspan="2"></td></tr>
<tr><td></td><td></td><td colspan="2"></td><td colspan="2"></td></tr>
<tr><td></td><td></td><td colspan="2"></td><td colspan="2"></td></tr>
<tr><td></td><td></td><td colspan="2"></td><td colspan="2"></td></tr>
<tr><td>Plan instructions</td><td colspan="5"></td></tr>
<tr><td>Class</td><td></td><td>Group No.</td><td></td><td>Group leader</td><td></td></tr>
<tr><td>Signature of teacher</td><td></td><td>Date</td><td colspan="3"></td></tr>
<tr><td colspan="6">Comments</td></tr>
</table>

II. Task Implementation Phase 1: Vegetable Seed Treatment and Seed Soaking for Germination

(I) Purpose of the Implementation of the Task Phase

To understand the role of pre-sowing treatment of vegetable seeds, and to master the methods of seed soaking and germination for vegetable seeds.

(II) Materials and Tools

Various types of seeds, culture dishes, filter paper, gauze, forceps, beakers, glass rods, thermometers, electric stoves, incubators, etc.

(III) Methods and Steps

Each group should perform the following operations according to the requirements.

1. Seed selection

Choose suitable vegetable varieties and inspect the seeds for maturity, plumpness, color, cleanliness, presence of pests and diseases, mechanical damage, germination potential, and germination rate, among other indicators.

2. Seed disinfection

(1) Agent treatment.

① Tomato seeds: a. Use a 100-fold dilution of 40% formalin solution and a 10% trisodium phosphate solution. Soak the seeds in clean water for 3–4 h. Then immerse the damp seeds in the solution for 15–20 min. Remove the seeds and wrap them in a wet cloth, sealing them in a container or bowl for 2–3 h. Afterward, rinse the seeds with clean water. This treatment can prevent early blight and viral diseases. b. Soak the seeds in a 200-fold dilution of hydrochloric acid for 3 hours to eliminate tomato tobacco mosaic virus. c. Soak the seeds in warm water at 40°C for 3–4 h, then transfer them to a 0.1% potassium permanganate solution for 30 min. Afterward, rinse them thoroughly with clean water. This treatment can reduce tomato canker and leaf spot disease.

② Eggplant seeds: a. Soak eggplant seeds in a 1% potassium permanganate solution for 30 min. Remove the seeds and rinse them thoroughly. Follow this with warm water soaking for germination. b. Soak the seeds in a 0.1% solution of carbendazim for 1 h, then wash them before initiating germination. This treatment is effective in preventing *Verticillium* wilt. c. Immerse the seeds in a 40% formalin solution for 15 min, then rinse them with clean water. This can help prevent eggplant brown spot disease and *Verticillium* wilt.

③ Pepper seeds: a. Soak pepper seeds in a 10% copper sulfate solution to prevent anthracnose and bacterial leaf spot. First, soak the seeds in clean water for 4–5 h, then immerse them in the copper sulfate solution for 5 min. Wash the seeds thoroughly before germination. b. Soak the seeds in a 0.2%–0.4% zinc sulfate solution at 20–25°C for 12 h to suppress the occurrence of pepper virus diseases. c. Soak the seeds in cold water for 3–4 h, then move them into a 500-fold dilution of 50% carbendazim solution for 1 h. Rinse the seeds with clean water. This treatment can kill or purify pepper viruses.

④ Melon seeds: There are two methods: powder mixing and liquid soaking. a. Powder mixing. After soaking in clean water, thoroughly mix the seeds with a 0.3% insecticide or fungicide. It can be directly mixed with dry seeds. Commonly used fungicides include Ridomil, Carbendazim, Thiram, and Tuzet. Insecticides such as Dipterex powder can also be used. b. Liquid soaking. The seeds are first soaked in clean water and then immersed in a liquid solution for a specified time to disinfect. Typically, the seeds can be soaked in a 100-fold dilution of 40% formalin solution for 30 min. To prevent fungal diseases, soak the seeds in a 500-fold dilution of 50% Carbendazim solution for 1 h, or in a 800-fold dilution of 72.2% Prochloraz water-based solution for 0.5 h. Then, soak the seeds in a 500-fold dilution of 50% Amobam solution for 1 h or in a 500-fold dilution of Imidazole Hydrochloride solution for 1–2 h. To prevent cucumber fusarium wilt, you can soak the seeds in a 2% – 3% bleach solution for 30–60 min. To prevent bacterial diseases, you can also use a 500-fold dilution of streptomycin sulfate solution with 1 million units for soaking the seeds for 2 h. To prevent viral diseases, you can soak the seeds in a 10% sodium phosphate solution for 20 min. After chemical treatment, it's essential to rinse the seeds thoroughly with clean water before germination or sowing.

⑤ Legume seeds. Soak them in a 0.1% copper sulfate solution for 15 min. After soaking, remove the seeds and rinse them thoroughly to remove any remaining solution on the seed surface. Alternatively, you can scald the seeds with hot water at 55°C for 5 min, then add cold water to bring the temperature to 25–28°C, soak the seeds for 3–4 h, and finally, remove the seeds for drying before sowing. Since legumes are sensitive to temperature and humidity fluctuations, they are typically not subjected to pre-germination before sowing.

(2) Hot water treatment. There are two methods for hot water treatment: warm water soaking and hot water scalding. a. Warm water soaking. Place the seeds in a clay pot and slowly pour 50–55°C warm water over them while stirring continuously. Continue stirring for 10–15 min until the water temperature drops to 30°C. Then, stop stirring and allow the seeds to soak. This method can be used for melons and nightshades. b. Hot water scalding. Start by immersing the seeds in cool water until they are just covered. Then, use hot water at 80–90°C while stirring until the water temperature reaches 55°C. Stop stirring and maintain this temperature for 7–8 min before proceeding with soaking. The amount of hot water should not exceed 5 times the quantity of seeds. This method is suitable for seeds with thicker seed coats, such as winter melon, eggplant, and cucumber seeds. During scalding, ensure that the seeds are thoroughly dried, as seeds with lower moisture content can better withstand high-temperature exposure.

(3) Seed soaking and germination.

① Seed soaking. Before germination, soak the seeds in water at 20–30°C. This allows the seeds to absorb moisture and speeds up the germination process. For seeds with easily sticking seed coats or those that haven't been properly fermented, such as cucumber, squash, and pumpkin seeds, you can first wash them with a 0.2%–0.4% alkaline solution. After soaking, remove the seeds, drain excess water, and then proceed with germination. The soaking time may vary depending on the types of seeds. For melon seeds, after soaking, use fine sand to rub off any sticky substance from the seed coats, separate the seeds with clean water, and rinse them thoroughly to prepare for germination.

② Germination. Vegetable seeds germinate most rapidly under conditions of suitable moisture, good aeration, and darkness at temperatures between 25°C and 30°C. It's important to control the germination temperature. Initially, keep it slightly lower, and gradually raise it. When the radicle is about to break through the seed coat, lower the temperature again to encourage robust radicle development. Turn the seeds every 4–5 h for adequate aeration and to ensure even temperature distribution. For seeds with longer germination periods, rinse them with warm water once a day to remove any mucilage and prevent mold growth. When around 75% of the seeds have cracked or roots have emerged, stop the germination process and prepare for sowing.

Some vegetable seeds, such as seedless watermelon, luffa, bitter gourd, have thick and hard seed coats, making germination difficult. By cracking the seed coat before soaking, you can improve seed germination rates and ensure uniform sprouting.

(IV) Task Record Sheet

Task name		Guiding teacher	
Group No.		Group leader	
Time		Record content	
Signature of teacher		Date	

(V) Assessment Sheet

Task name	Vegetable seed treatment and seed soaking for germination				Group No.		
Implementation date		Process record for vegetable seed treatment and seed soaking for germination, ________ pages in total					
Evaluation item	Evaluation content		Percentage	Teacher's mark	Student's mark	Score	Total
Process evaluation	Working attitude	Work attendance	2%	1%	1%		
		Responsible attitude	3%	2%	1%		
		Communication	2%	1%	1%		
		Teamwork	3%	2%	1%		
	Working method	Learning ability	3%	8%	2%		
		Planning ability	3%	2%	1%		
		Problem-solving ability	4%	3%	1%		
	Practical operation	Seed disinfection	5%	3%	2%		
		Quality of selected seeds	3%	2%	1%		
		Selection of seed soaking container and water quantity	5%	4%	1%		
		Number of times and method of seed washing	5%	4%	1%		
		Control of soaking depth and time	7%	5%	2%		

continued

Evaluation item	Evaluation content		Percentage	Teacher's mark	Student's mark	Score	Total
Process evaluation	Practical operation	Reasonable germination temperature setting	7%	5%	2%		
		Washing and turning during germination	3%	2%	1%		
		Timing for stopping germination	5%	4%	1%		
Result evaluation	Result of seed soaking and germination	Effect of seed soaking and germination	10%	8%	2%		
		Analyzing the rationality of soaking and germination methods	10%	8%	2%		
	Training report	Correctness and compliance of report filling	20%	16%	4%		

III. Task Implementation Phase 2: Vegetable Seed Sowing and Seedling Raising

(I) Purpose of the Implementation of the Task Phase

By raising seedlings of vegetables, you can understand the seedling raising process and master the key techniques for preparing seedling beds and sowing vegetables.

(II) Materials and Tools

Greenhouse, vegetable seeds, garden soil, organic fertilizer, etc.

(III) Methods and Steps

Each group should perform the following operations according to the requirements.

1. Vegetable seedling bed preparation

(1) Nursery soil preparation: The specific formula for nursery soil can vary depending on different vegetables and the seedling stages. Currently, a common formula for seedling beds is 6 parts topsoil and 4 parts well-rotted organic fertilizer. After sifting garden soil and organic fertilizer, mix in quick-acting fertilizer thoroughly and allow it to sit overnight.

(2) Seedling bed preparation: Use suitable seedling-raising facilities, and once the facilities are ready, lay out the nursery bed. The bed should have a width of 1–1.5 m. Spread the nursery soil evenly on the bed, with a thickness of about 10 cm, and then level the bed surface.

2. Sowing

(1) Pre-sowing preparation: Based on the chosen vegetable type, determine the appropriate sowing time and quantity and treat the seeds accordingly.

(2) Sowing: In cold seasons, it's best to sow on a warm morning. Before sowing, thoroughly water the bed. After the water has soaked in, sprinkle a thin layer of nursery soil on the bed surface. Seeds of melons, tomatoes, cabbage, and leafy greens are generally broadcast, while legume seeds are usually sown in holes. Seeds treated for germination are moist on the surface and not easily dispersed; They should be mixed with fine sand or herbal ash before sowing. After sowing, cover the seeds with soil and then use a thin film to cover the bed surface.

3. Post-sowing management

Observe the emergence of seedlings daily and keep records while enhancing seedling care during this period.

(IV) Task Record Sheet

Task name		Guiding teacher	
Group No.		Group leader	
Time		Record content	

Signature of teacher		Date	

(V) Assessment Sheet

<table>
<tr><td>Task name</td><td colspan="4">Vegetable seed sowing and seedling raising</td><td>Group No.</td><td colspan="2"></td></tr>
<tr><td>Implementation date</td><td></td><td colspan="6">Process record for vegetable seed sowing and seedling raising, ________ pages in total</td></tr>
<tr><td>Evaluation item</td><td colspan="2">Evaluation content</td><td>Percentage</td><td>Teacher's mark</td><td>Student's mark</td><td>Score</td><td>Total</td></tr>
<tr><td rowspan="8">Process evaluation</td><td rowspan="4">Working attitude</td><td>Work attendance</td><td>2%</td><td>1%</td><td>1%</td><td></td><td></td></tr>
<tr><td>Responsible attitude</td><td>3%</td><td>2%</td><td>1%</td><td></td><td></td></tr>
<tr><td>Communication</td><td>2%</td><td>1%</td><td>1%</td><td></td><td></td></tr>
<tr><td>Teamwork</td><td>3%</td><td>2%</td><td>1%</td><td></td><td></td></tr>
<tr><td rowspan="3">Working method</td><td>Learning ability</td><td>3%</td><td>1%</td><td>2%</td><td></td><td></td></tr>
<tr><td>Planning ability</td><td>3%</td><td>2%</td><td>1%</td><td></td><td></td></tr>
<tr><td>Problem-solving ability</td><td>4%</td><td>3%</td><td>1%</td><td></td><td></td></tr>
<tr><td>Practical operation</td><td>Completeness of the preparation work</td><td>2%</td><td>1.5%</td><td>0.5%</td><td></td><td></td></tr>
</table>

continued

Evaluation item	Evaluation content		Percentage	Teacher's mark	Student's mark	Score	Total
Process evaluation	Practical operation	Rationality of preparing the bed soil	3%	2.5%	0.5%		
		Rationality of choosing disinfectants	2%	1.5%	0.5%		
		Accuracy of seed quantity calculation	3%	2.5%	0.5%		
		Quality of seedling bed preparation	5%	4%	1%		
		Thoroughly watering the bottom	7%	6%	1%		
		Uniformly and timely sowing	8%	5%	3%		
		Even and reasonable soil covering thickness	5%	4%	1%		
		Seeding proficiency	5%	4%	1%		
Result evaluation	Sowing and seedling raising result	Emergence quality	10%	8%	2%		
		Analyzing the rationality of sowing methods	10%	8%	2%		
	Training report	Correctness and compliance of report filling	20%	16%	4%		

IV. Task Implementation Phase 3: Sowing of Tuberous Vegetables

(I) Purpose of the Implementation of the Task Phase

Understand the seeding techniques for tuberous vegetables by selecting seed tubers, treating them, and planting them.

(II) Materials and Tools

Greenhouse, seed tubers of tuberous vegetables, various disinfectants, etc.

(III) Methods and Steps

Each group should perform the following operations according to the requirements.

1. Seed potato treatment

For potatoes, about 150 kg of seed potatoes are needed per mu. 20–25 days before planting, place the seed potatoes in a warm, sunny area to air them out for 2–3 days. During this time, remove any diseased or rotten potatoes. Then, proceed to cut the seed potatoes into pieces. When cutting, make the most of the apical dominance, and cut them diagonally towards the top in a spiral fashion. After that, divide them into halves or quarters according to the terminal bud, with each seed potato piece having 1–2 sprout eyes and weighing 25–30 g. Apply 50 mL of iprodione and 20 mL of imidacloprid per 100 kg of seed potatoes. After drying the cut surfaces, place them in a room with a temperature of 18 – 20°C and use a layered method for sprouting. Once the sprouts reach about 3 cm in length, expose them to diffused light for greening and thickening before planting.

For taros, choose daughter taros that are free of disease, pests, and mold, have full terminal buds, and uniform corms. After air-drying for 2–3 days, you can proceed with sprouting and planting.

Ginger should be in single pieces weighing 50–70 g each. About 25–30 days before planting, place them in a warm, temperature-controlled room or an environment at 20°C. Air-dry them for 1–2 days. Afterwards, soak them in a 200-fold dilution of potassium permanganate solution for 10–20 min and air-dry them again. Sprouting can be done in a greenhouse or warm environment at a temperature of 22–25°C, which usually takes around 25 days for the young sprouts to reach about 1–1.5 cm in length.

2. Seeding

The planting density for potatoes varies depending on the varieties. For the Atlantic variety, you should plant at a density of 5.25–6.00 thousand plants per hm^2 with a spacing of 24–26 cm between plants. It's recommended to use 2,250–2,700 kg of seed potatoes per hm^2, and the planting depth should be approximately 10 cm. After planting, you can use 1,500–1,950 mL/hm^2 of acetochlor with 450–600 kg/hm^2 of water for spraying to eliminate weed seedlings.

Ginger should be sown on clear, warm days, and it should be covered by a canopy 4–5 days before planting. The planting density for ginger is 5,500–6,000 plants per mu. Rows should be spaced 55–60 cm apart, and the distance between plants in a row should be 18–23 cm. Plant the ginger at a depth not exceeding 7 cm below ground level, with a covering thickness of 2–3 cm. It's important to water the furrows before sowing, allowing the water to seep in, and then place the ginger evenly and neatly with the young shoots facing in the same direction.

For taro planting, the plant spacing should be about 40 cm. Taro is typically planted in grades of different sizes. Before sowing, make sure to thoroughly water the soil. After covering the seeds with soil, you can apply pre-emergent weed control spray.

(IV) Task Record Sheet

Task name		Guiding teacher	
Group No.		Group leader	
Time		Record content	
Signature of teacher		Date	

(V) Assessment Sheet

Task name	Direct sowing of tuberous vegetables				Group No.		
Implementation date		Process record for direct sowing of tuberous vegetables, _______ pages in total					
Evaluation item	Evaluation content		Percentage	Teacher's mark	Student's mark	Score	Total
Process evaluation	Working attitude	Work attendance	2%	1%	1%		
		Responsible attitude	3%	2%	1%		
		Communication	2%	1%	1%		
		Teamwork	3%	2%	1%		
	Working method	Learning ability	3%	1%	2%		
		Planning ability	3%	2%	1%		
		Problem-solving ability	4%	3%	1%		
	Practical operation	Completeness of the preparation work	2%	1.5%	0.5%		
		Rationality of choosing disinfectants	4%	3%	1%		
		Accuracy of seed quantity calculation	8%	6%	2%		
		Thoroughly watering the bottom	6%	4%	2%		
		Uniformly and timely sowing	10%	8%	2%		
		Even and reasonable soil covering thickness	10%	6%	4%		

continued

Evaluation item	Evaluation content		Percentage	Teacher's mark	Student's mark	Score	Total
Result evaluation	Sowing and seedling raising result	Emergence quality	10%	8%	2%		
		Analyzing the rationality of sowing methods	10%	8%	2%		
	Training report	Correctness and compliance of report filling	20%	16%	4%		

V. Task Implementation Phase 4: Land Preparation, Mulching Film Covering, Transplanting

(I) Purpose of the Implementation of the Task Phase

By preparing the vegetable field, using mulching film, and transplanting, you will learn the key techniques for vegetable plot preparation, mulching film application, and transplanting.

(II) Materials and Tools

Hoe, mulching film, vegetable seedlings, etc.

(III) Methods and Steps

Each group should perform the following operations according to the requirements.

1. Soil preparation and ridging

(1) Solanaceous vegetables.

① Raised bed and furrow planting: The bed width is 80 cm, the bed height is 20–25 cm, and the distance between the centers of two planting furrows is 50 cm. The planting furrow entrance is 15–20 cm wide, and the bottom is 12–15 cm wide. ② High ridge and furrow planting: The ridge spacing is 60–65 cm, the ridge height is 20–25 cm, the planting furrow is 20 cm deep, the entrance is 15–20 cm wide, and the bottom is 15 cm wide. ③ Apply sufficient base fertilizer: 7–10 days before planting, apply over 2,000 kg of well-rotted manure per mu, and add 30–50 kg of compound fertilizer. Apply the fertilizer deep in the ridges (furrows) to meet the needs of growth and yield.

(2) Legumes.

Legumes should not be continuously cultivated in the same plot. For the previous crop, it's best to choose cabbage, scallions, garlic, or a plot that hasn't grown legumes for 2–3 years, preferably fallow land. Prepare the soil early and deeply, promote soil maturation. Typically, per mu, apply over 3,000 kg of well-rotted manure, 30–40 kg of calcium phosphate, 50–75 kg of straw ash or bran ash, or 10–20 kg of potassium sulfate as base fertilizer. For acidic soils, apply lime appropriately. Create high ridges with a width of 133 cm (including furrows) and dig deep furrows.

(3) Cabbage vegetables.

Seedling bed preparation: The seedling bed should have a raised bed width of 80 cm, a bed width of approximately 1.6 m, a length of 8–12 m, and a height of around 13 cm. The two planting furrows should be centered 10 cm apart, with furrow depths of 0.8–1 cm.

(4) Root vegetables.

It is recommended to avoid planting radishes after cruciferous vegetables and nightshade vegetables such as tomatoes and peppers. It's preferable to choose melons or leguminous vegetables as the previous crops. Land preparation should be meticulous, involving early and thorough plowing. Deep plowing combined with the application of basal fertilizer is crucial. For every mu, apply 2,500–5,000 kg of well-rotted farmyard manure, thoroughly mixed with the soil. In northern regions, raised beds are commonly used with bed heights of 20–30 cm and bed spacing of 40–60 cm, or flat beds with a width of 1.2–1.5 m. In southern regions, deep furrows with raised beds are often used with bed widths of 1.5 m (including furrows) and bed heights of 15–20 cm.

(5) Leafy vegetables.

For celtuce, you need a significant amount of fertilizer. Choose well-fertilized and loose loamy soil. Apply 3,000–4,000 kg of manure or garbage per mu, and 1,500–2,000 kg of human manure as the base fertilizer. After turning the soil, create beds with a width of 1.6 m (including furrows). Celery prefers fertile and loose soil. After the previous crop is removed, immediately clear away crop residues and weeds and ensure proper water channels. Create beds with a width of 1 m. Before forming the beds, apply 5,000 kg or more of high-quality farmyard manure, 100 kg of ammonium dihydrogen phosphate, 100 kg of wood ash, and 10 kg of urea per mu. Deeply plow the soil to a depth of 20–30 cm, ensuring thorough mixing of the fertilizer and soil. After fine harrowing, the land is ready for planting.

(6) Allium vegetables.

For the root system of allium vegetables, it's important to choose sandy loam soil that is fertile and rich in organic matter. These vegetables have relatively weak water and nutrient absorption abilities through their roots. Avoid clayey soil as it hinders root development and bulb expansion, and sandy soil may have poor water and nutrient retention. Onions should not be planted continuously in the same soil. It is best to choose a preceding crop that received a relatively high amount of fertilization, such as nightshades, melons, or legume vegetables. Apply 2,000–3,000 kg of well-rotted farmyard manure per mu, along with 30 kg of calcium superphosphate and compound fertilizer. Plow the soil to a depth of 20 cm, harrow it, and create ridges. In the Yangtze River basin and southern regions, create deep furrowed ridges with a width of 1.5–2 m, a ridge-groove width of 40 cm, and a groove depth of about 20 cm for better drainage. In northern regions, create flat ridges with a width of 1.6–1.7 m.

(7) Tuberous vegetables.

Selecting the right land is crucial, and it's important to avoid planting the same crop in the same field continuously. Choose soil that is loose, deep, and well-draining, with medium to high fertility levels. After harvesting late rice, perform deep plowing to a depth of 25–30 cm. As part of the deep plowing process, apply 50 kg of hydrated lime per mu for disinfection. Before planting, make sure the soil is well-pulverized and leveled. Create raised beds with a width of 1.2 m, a trench width of 0.4 m, and a depth of at least 0.15 m. Implement appropriate drainage channels, such as perimeter and cross trenches, to prevent waterlogging. Apply 1,500–2,000 kg of compost or 750 kg of well-rotted chicken manure per mu, along with 50–100 kg of organic fertilizer, mixed and applied in the planting rows. Apply potassium sulfate compound fertilizer between the rows.

2. Mulching film covering

The method of covering with mulching film: After spraying the herbicide, the plastic film should be immediately covered. When manually covering the film, it's recommended to have at least three people in a group. Start by burying one end of the mulching film at the beginning of the ridge or bed, pressing it down firmly. One person should then unfold the mulching film, while two others work on both sides of the ridge or bed to cover the film's edges with soil, ensuring that the mulching film is pulled tight, laid evenly, and tightly sealed against the ridge or bed surface. The covered area, which is the width of the transparent part, should account for 3/5 of the ridge or bed surface. The remaining plastic film should flow into the furrow for field operations and irrigation.

3. Transplanting

(1) Solanaceous vegetables and cabbages.

For the transplanting date, it is important to pay attention to the weather forecast and schedule the transplanting day to be at the end of cold weather and the beginning of warm weather, which is conducive to seedling survival. Planting generally begins around midday, and it's advisable to concentrate efforts and complete the transplanting work within the greenhouse before the afternoon. The day before transplanting, the seedling bed should be watered adequately to ensure that there is as much soil attached to the seedlings as possible during transplanting, reducing damage to the seedling root system. Additionally, a pesticide treatment should be conducted before transplanting, typically using fungicides such as carbendazim or chlorothalonil. After breaking the mulch cover, plant two rows in each raised bed or ridge. For tomatoes, plant 2,500–3,000 plants per mu; for chili peppers, 3,000 plants; for eggplants, 2,200–2,400 plants per mu. Maintain the planting depth at the level of the seedbed. After planting, apply well-rotted liquid fertilizer to establish strong roots.

(2) Legumes.

The ideal time for transplanting is when the first pair of true leaves have fully developed, and the first compound leaf is unfurling. It's recommended to transplant seedlings that are 20–25 days old. The transplanting depth should be such that the paper pots are completely buried in the soil. During transplanting, it's important to water the seedlings to ensure a smooth transition.

(3) Root vegetables.

After breaking the mulch cover, plant two rows in each raised bed or ridge. For hole planting, sow 5–6 seeds per hole for Beishan Radish and 2–4 seeds per hole for varieties like Hefeng, Xiakang 40, and Taibai. Use a seeding rate of 100–150 g per mu, with rows spaced 40 cm apart and holes 20 cm apart.

(4) Leafy vegetables.

The planting density for lettuce varies depending on the varieties and seasons. For early-maturing varieties, the row and plant spacing should be approximately 18-23 cm^2, with 8,500–12,000 plants per mu. For late-maturing varieties, the row and plant spacing should be around 25–30 m^2, with 7,000–8,000 plants per mu.

Celery is often planted as single plants, but in some areas, 1–2 or 2–3 plants per hole may be used. The row spacing varies depending on the varieties, generally 10–20 cm, with a plant spacing of around 10 cm. For larger varieties like Western celery, a row spacing of 16–20 cm^2 is suitable, with a density of 37,000–42,000 plants per mu. On the day before planting, water the seedbed. When lifting seedlings, dig them up with their roots, remove soil, and eliminate weak or diseased seedlings. When planting, grade the seedlings by size and plant similarly sized ones together. Use a pointed shovel to dig deep holes when planting, ensuring that the young seedlings' root systems are inserted into the soil. The planting depth should match the depth at which the seedlings were grown in the nursery. Immediately after planting, water them thoroughly.

(5) Allium vegetables.

Onions are ideally planted in early November. When planting, it's important to grade the seedlings, first planting the standard-sized seedlings and then the smaller ones. The planting depth should be appropriate: planting too deep can lead to spindle-shaped bulbs with lower yields, while planting too shallow can cause lodging. It's recommended to bury the bulbs about 1 cm deep. For each mu of planting area, apply ample well-rotted organic fertilizer (4,000 kg), ammonium dihydrogen phosphate (50 kg), and potassium sulfate (30 kg) before planting. After fertilization, level and finely rake the soil, and create flat beds with a width determined by the width of the mulching film. Plants with a spacing of 20 cm × 15 cm, typically resulting in around 23,000 plants per mu. Before covering them with mulching films, water the soil. Then, make holes in the plastic mulch with a depth of 3 cm and a diameter of 1.5 cm. During planting, insert the seedlings directly into these holes, press them down with your hands, and seal the hole. If you're not using mulching films or haven't watered the soil before covering the plants, be sure to water them immediately after planting and provide a light watering 4–5 days later.

The planting time for garlic chives is in the mid- to late August. Before planting, you should lift the seedlings and shake off the soil. Grade the seedlings based on their size, cut them uniformly, and prepare them for transplanting. The planting method typically involves using a furrow planting technique. Here's the specific procedure: dig furrows with a depth of 10–15 cm and plant each clump of garlic chives every 25 cm along the furrow. After placing them, cover them with soil to a depth of 4 cm and then water to ensure that the roots are in close contact with the soil. About 5–6 days after transplanting, water the chives once more to promote their growth.

For garlic, you should prepare a raised bed with a width of 140–160 cm and a length of 1,000–1,500 cm oriented north to south. Break up any clods and level the bed. Within the raised bed, create furrows with row spacing of 20 cm and a depth of 3–4 cm. In these furrows, plant the garlic cloves with a spacing of 8 cm between each clove. This planting density results in 40,000 plants per mu, and you will need 200 kg of garlic seeds per mu.

(IV) Task record Sheet

Task name		Guiding teacher	
Group No.		Group leader	
Time		Record content	
Signature of teacher		Date	

(V) Assessment Sheet

<table>
<tr><td>Task name</td><td colspan="5">Land preparation, mulching film covering, transplanting</td><td>Group No.</td><td colspan="2"></td></tr>
<tr><td>Implementation date</td><td></td><td colspan="7">Process record for land preparation, mulching film covering, transplanting, ________ pages in total</td></tr>
<tr><td>Evaluation item</td><td colspan="2">Evaluation content</td><td>Percentage</td><td>Teacher's mark</td><td>Student's mark</td><td>Score</td><td>Total</td></tr>
<tr><td rowspan="7">Process evaluation</td><td rowspan="4">Working attitude</td><td>Work attendance</td><td>2%</td><td>1%</td><td>1%</td><td></td><td></td></tr>
<tr><td>Responsible attitude</td><td>3%</td><td>2%</td><td>1%</td><td></td><td></td></tr>
<tr><td>Communication</td><td>2%</td><td>1%</td><td>1%</td><td></td><td></td></tr>
<tr><td>Teamwork</td><td>3%</td><td>2%</td><td>1%</td><td></td><td></td></tr>
<tr><td rowspan="3">Working method</td><td>Learning ability</td><td>3%</td><td>8%</td><td>2%</td><td></td><td></td></tr>
<tr><td>Planning ability</td><td>3%</td><td>2%</td><td>1%</td><td></td><td></td></tr>
<tr><td>Problem-solving ability</td><td>4%</td><td>3%</td><td>1%</td><td></td><td></td></tr>
</table>

continued

Evaluation item	Evaluation content		Percentage	Teacher's mark	Student's mark	Score	Total
Process evaluation	Practical operation	Completeness of the preparation work	2%	1.5%	0.5%		
		Degree of soil preparation	3%	2%	1%		
		Quality of mulching film covering	5%	4%	1%		
		Rationality of the transplanting time	5%	4%	1%		
		Rationality of the use of base fertilizer	3%	2%	1%		
		Transplanting depth	2%	1.5%	0.5%		
		Transplanting density	5%	3%	2%		
		Seedling growth condition	5%	4%	1%		
		Transplanting proficiency level	10%	8%	2%		
Result evaluation	Transplanting result	Seedling survival rate and quality	10%	8%	2%		
		Analyzing the rationality of transplanting methods	10%	8%	2%		
	Training report	Correctness and compliance of report filling	20%	16%	4%		

VI. Task Implementation Phase 5: Grafting Seedling Technology for Melons

(I) Purpose of the Implementation of the Task Phase

Grafting seedlings for vegetables is an effective method to prevent soil-borne diseases and enhance plant resistance. It is a fundamental technique in modern greenhouse horticulture and finds extensive use in the cultivation of cucurbits and solanaceous crops. In this experiment, we explore the grafting process for cucurbitaceous seedlings and the subsequent care of grafted seedlings. We aim to understand the application of grafting technology in cucurbitaceous vegetable crops and the factors influencing grafting success rates. Additionally, we seek to master the common grafting methods employed in cucurbitaceous vegetable cultivation.

(II) Materials and Tools

(1) Materials: well-cultivated melon scion seedlings and rootstock seedlings.

(2) Tools: knife blades, bamboo sticks, gauze or alcohol swabs, plastic clips, small spray bottles, etc.

The grafting site should not have direct sunlight and should have an air humidity of over 80%.

(III) Methods and Steps

Each group should perform the following operations according to the requirements.

1. Pre-grafting preparation

(1) Preparation of scions and rootstock seedlings: Grafted cucumber seedlings are generally grafted onto black-seeded pumpkin rootstocks. For the splice grafting method, rootstock seeds should be sown 3–5 days earlier than scions, while for the approach grafting method, rootstock seeds should be sown 3–5 days later than scions. The methods for seed treatment, sowing, and management are the same as conventional practices. Grafting can be done when the seedlings have grown to the stage just before the first true leaf unfolds. Specific requirements are as follows: For splice grafting, rootstock seedlings

should have sturdy hypocotyls, and grafting should be done before the first true leaf unfolds on the scion. For approach grafting, scions should be slightly longer than the hypocotyls of the rootstock.

(2) Preparation of grafting seedbed: Use a warming bed or a small greenhouse with good humidity and temperature control. Water the seedbed thoroughly one day before grafting.

2. Selection of grafting method

There are three basic grafting methods for cucurbits: cleft grafting, splice grafting, and approach grafting. Cleft grafting is less commonly used because it can cause grafting union failure due to the uneven development of vascular bundles at the graft junction, affecting the growth of grafted seedlings. Approach grafting is characterized by high grafting success rates and simple management during the grafting process, and is particularly effective in early spring during low-temperature conditions and in the hot summer and fall seasons when grafting success rates are higher. Splice grafting is a grafting method that offers a balance between ease of mastering the technique, high efficiency, and high grafting success rates. This method is widely used but requires strict management during the grafting process.

(1) Splice grafting method.

① Preparation of the rootstock: First, remove the rootstock's growing point with a blade. Then, take a bamboo stick with one end gradually tapered and of an appropriate thickness matching the lower stem of the scion. Insert the bamboo stick into the cut surface of the rootstock from the side where the removed growing point used to be, with one side of the cotyledon facing downward, creating a hole approximately 1 cm deep. The depth should not pierce the epidermis of the hypocotyl, and the bamboo stick should be faintly visible.

② Preparation of the scion: Remove the scion seedling and use a blade to make a slanting cut approximately 0.5 cm below the growing point, shaping it into a wedge about 1 cm long. Remove the bamboo stick and promptly insert the prepared scion into the hole in the rootstock. Ensure that the cotyledons of the rootstock and scion fit closely together, forming a cross shape with the cotyledons of both the rootstock and scion. After grafting, fix it in place with a grafting clip.

(2) Approach grafting method.

① Preparation of the rootstock: Use a bamboo stick to take both types of seedlings from the seedbed. Begin by removing the growing point of the rootstock seedling. From approximately 1 cm below the cotyledons, make a downward cut at a 30° angle, cutting to a depth that is half the thickness of the stem but no more than two-thirds. After the cut, hold the rootstock gently in your left hand.

② Preparation of the scion: Take the scion seedling and cut it from about 2.5 cm below the cotyledons at a 30° angle, cutting halfway through. Then, match the two seedlings so that their cut surfaces fit together, immediately secure them with a grafting clip, and then plant them in the grafting seedbed.

3. Notes on grafting

(1) After taking out the seedlings, rinse off the soil from the root system with clean water.

(2) Graft quickly and ensure that the cut surfaces fit accurately, with the clip securing the side of the cantaloupe stem.

(3) Cover the grafted seedlings with a row cover to maintain warmth, humidity, and shade. Make sure that no soil comes into contact with the grafting site.

(4) Wet the entire seedbed surface, but avoid spraying water to prevent water from entering the grafting site. Do not bury the grafting clips. Gently press both roots of the grafted seedling into the soil and fill with additional soil.

(5) Maintain aseptic conditions throughout the grafting process, and ensure that the grafted area does not come into contact with the soil.

(6) When planting the grafted seedlings, do not bury them too deep to prevent the graft site from developing roots. Position the root system of the stock plant in the center of the planting hole, and then gently place the scion root system to the side, covering it with a little soil.

(IV) Task Record Sheet

Task name		Guiding teacher	
Group No.		Group leader	
Time		Record content	
Signature of teacher		Date	

(V) Assessment Sheet

<table>
<tr><td>Task name</td><td colspan="5">Grafting seedling for melon vegetables</td><td>Group No.</td><td colspan="2"></td></tr>
<tr><td>Implementation date</td><td></td><td colspan="7">Process record for melon vegetable seed sowing and seedling raising, ________ pages in total</td></tr>
<tr><td>Evaluation item</td><td colspan="2">Evaluation content</td><td>Percentage</td><td>Teacher's mark</td><td>Student's mark</td><td>Score</td><td>Total</td></tr>
<tr><td rowspan="4">Process evaluation</td><td rowspan="4">Working attitude</td><td>Work attendance</td><td>2%</td><td>1%</td><td>1%</td><td></td><td></td></tr>
<tr><td>Responsible attitude</td><td>3%</td><td>2%</td><td>1%</td><td></td><td></td></tr>
<tr><td>Communication</td><td>2%</td><td>1%</td><td>1%</td><td></td><td></td></tr>
<tr><td>Teamwork</td><td>3%</td><td>2%</td><td>1%</td><td></td><td></td></tr>
</table>

continued

Evaluation item	Evaluation content		Percentage	Teacher's mark	Student's mark	Score	Total
Process evaluation	Working method	Learning ability	3%	1%	2%		
		Planning ability	3%	2%	1%		
		Problem-solving ability	4%	3%	1%		
	Practical operation	Grafting time	10%	8%	2%		
		Disinfection of grafting tools	5%	4%	1%		
		Grafting operations' rationality	15%	10%	5%		
		Management after grafting	5%	4%	1%		
		Pruning roots and removing suckers	5%	4%	1%		
Result evaluation	Sowing and seedling raising result	Grafting success rate	10%	8%	2%		
		Grafting seedling quality	10%	8%	2%		
	Training report	Correctness and compliance of report filling	20%	16%	4%		

VII. Task Implementation Phase 6: Adjustment of Melon and Solanaceous Vegetable Plants

(I) Purpose of the Implementation of the Task Phase

Understand the impact of plant pruning on the growth, development, yield, and quality of vegetables, and master plant adjustment techniques in cucurbit and solanaceous vegetable production, including vine training, trellising, pruning, leaf and fruit thinning, and stem pruning.

(II) Materials and Tools

Bamboo, nylon rope, 12-gauge wire, scissors

(III) Methods and Steps

1. Observe plant types and plant growth status

2. Cucumber plant pruning

When cucumber seedlings reach 6–7 leaves, it's important to promptly trellis them and remove lateral branches, tendrils, and staminate flower buds to reduce nutrient consumption. Approximately 15 days after transplanting, when cucumber seedlings reach the seventh leaf, string training can be started to suspend the vines. Start by stretching a wire horizontally, 20 cm above the cucumber rows, along the length of the greenhouse. Fasten it securely, running in a north-south direction. Tie a piece of polypropylene plastic rope or hemp twine above each cucumber plant. Secure the upper end of the rope with a dead knot, connecting it to the wire. Fasten the other end with a live knot at the base of the cucumber plant, allowing the vines to climb the rope, forming an "S" shaped trellising. During the vine training process, pay attention to adjusting the height of the vine tips. When the plant growth becomes excessively vigorous, you can gently guide the growing point away from the trellis wire, allowing it to naturally hang downwards. This helps to reduce the dominance of the apical growth. Conversely, you can encourage the stem tips to grow vertically along the rope, stimulating growth. Arrange the stem tip growing points of cucumber vines in a slanting line from north to south, ensuring even exposure to light, no shading, and uniform, orderly growth. After cucumber transplanting, during the plant's growth, twist the vines around the support wire as they grow one section at a time. Simultaneously, promptly remove staminate flowers, tendrils, and deformed cucumbers to conserve nutrients. In situations with high planting density, it's advisable to promptly remove lateral branches during the fruiting period. In the case of lower density or less dense foliage, you can retain more than 5 lateral branches, each bearing one cucumber. Additionally, leave 2 leaves before each cucumber and remove the growing tip after harvesting the cucumber. In the case of cucumber cultivation in a sunlit greenhouse, the focus is primarily on fruiting along the main vine. Typically, no pruning of the growing tip is done throughout the entire growth cycle. When there's an adequate nutrient supply, cucumbers can be grown backward on the main vine, allowing the main vine to reach heights of over 5 m or more. Therefore, when the leading end of the main vine approaches the roof, it's important to initiate vine dropping or trellis management to prevent overgrowth. Before initiating vine dropping, it's advisable to remove the lower, older leaves. Typically, in the rear section of a sunlit greenhouse, vine dropping is performed 2–3 times, while in the front section, it is carried out 3–4 times to manage the plant's growth. The method of falling a vine is to untie the nylon rope tied to the iron wire and let the lower old vine lie on the ground, leaving room for the dragon head to continue growing.

3. Cantaloupe plant pruning

For trellis cultivation of cantaloupes, it's essential to promptly install support lines once tendrils start to form. Nylon cord is a cost-effective and practical choice for this purpose, making it easy to manage the cantaloupe plants. To set up the support lines: First, install a 12-gauge wire about 30–40 cm below the greenhouse roof, running parallel to the rows of cantaloupe plants. Then, on the ground, parallel to the rows of cantaloupe plants, install another 12-gauge wire. Attach the upper end of the nylon cord to the wire on the greenhouse roof and allow the lower end to hang naturally directly above the cantaloupe plants. Secure it tightly to the lower wire. Finally, adjust the tension of the nylon cord to keep it slightly loose, ensuring it's not too tight, which facilitates the wrapping of tendrils.

When wrapping tendrils, ensure that all cantaloupe plants are wrapped in the same direction (either clockwise or counterclockwise) along the nylon cord. This promotes even the distribution of leaves, facilitating photosynthesis and the production, transport, and accumulation of sugars. While wrapping the tendrils, also remove any cotyledons, tendrils, or old leaves from the cantaloupe plants. This enhances ventilation, reduces the occurrence of diseases, and ensures sufficient light penetration. As the cantaloupe plants grow, continue to wrap the tendrils promptly to prevent the vine tips from drooping or hanging upside down, which can hinder normal growth.

To manage cantaloupe plants effectively, perform topping when 4–5 leaves are grown and implement double-vine pruning. Pruning activities such as removing lateral branches, pinching off shoots, and shoot tip removal should be carried out on sunny days, ensuring that wounds are dry by evening, as this prevents the plants from being susceptible to vine wilt disease. For plants with fewer than 10 nodes, remove lateral branches as early as possible. For plants with 11–15 nodes on the main vine, leave 2 leaves when pinching off the growing tips. For plants with more than 15 nodes, remove lateral branches entirely, and on the main vine, perform shoot tip removal at the 25th node. At the top, leave 2–3 lateral branches as fruit-bearing branches.

4. Watermelon plant pruning

Around 20 days after transplanting, watermelon plants begin to produce runners. At this stage, the external temperature is still relatively low, so the runners should continue growing inside the greenhouse. To prevent the growing points of the plants from getting too close to the greenhouse film and becoming scorched, it's advisable to guide the runners appropriately inside the greenhouse. For double-cover planting systems, where two plants are grown side by side, it is common to employ double-vine pruning. In single rows with two plants, one plant's vines are guided in one direction, and the other plant's vines are guided in the opposite direction. In double-row planting systems, one row of plants' vines is guided towards the other row of plants. When the runners are about 30 cm long, it's time to begin vine compression. At this stage, the watermelon runners are spread out on the mulching film. To compress the vines, use soil blocks, pressing the runners into the soil. For runners that have entered the vine extension ridges, use 6–7-cm-long, 0.5-cm-thick wooden sticks bent into a "V" shape to secure and compress the runners. In double-cover planting systems, where grafted seedlings are commonly used for transplanting, it's essential to avoid shallow trench compression to prevent the compression nodes from developing adventitious roots, which can reduce or even negate the disease resistance provided by grafting. Typically, it's recommended to compress one runner every four nodes. When the fruit sets, compress the two nodes before and after each fruit to protect against wind damage to the young fruit.

5. Winter melon plant pruning

Due to the ground-based cultivation method of winter melon, where plants sprawl along the ground, the planting density for winter melon is relatively low, with only 180 plants per mu. It's important to promptly adjust the growth direction of winter melon stems and vines to ensure that they evenly cover the entire bed surface. Additionally, when the stems and vines of winter melon plants extend beyond their designated bed surface, manual adjustments should be made to ensure that they stay within their allotted area. This is necessary to maintain proper growth of the winter melon plants.

6. Tomato plant pruning

For indeterminate tomato varieties, pruning is essential and can be done in two ways: single-stem pruning and double-stem pruning. Approximately 40–50 days before the final fruit harvest, it's important to remove the growing tips, a process known as "topping", while leaving 1–2 leaves in the upper part of the fruit cluster. Additionally, promptly remove old, diseased, and yellowing leaves at the base of the plant. For large and medium-sized fruit varieties with excessive flower clusters, you can thin the flowers or fruits during the flowering or early fruiting stage. Leave 3–4 fruits per cluster and remove any deformed fruits early in their development.

7. Chili plant pruning

If chili plants exhibit excessive vegetative growth and poor fruiting, you can perform various pruning techniques to adjust their growth: After harvesting the main pepper crop, remove all old leaves below the first lateral branch to improve ventilation and airflow. For plants with dense upper foliage, consider pruning away some inward-growing and weaker lateral branches between rows. Later in the autumn, a comprehensive plant adjustment will be performed. Remove all old leaves below the third lateral branch and trim some of the excessively long branches on the inner side of the plant. In cases where plants grow vigorously, become tall, and have fragile branches, you can use horizontal bamboo stakes on the outside of the bed to support the plants, preventing them from lodging.

8. Eggplant plant pruning

When eggplant plants reach a certain number of leaves, the terminal bud develops into a floral bud, and two adjacent leaf buds below the growing point produce lateral branches. These two lateral branches grow almost equally, effectively forming a double-stem branching structure to replace the main stem. Below the first lateral branch on the main stem, each leaf bud has the potential to develop into a lateral branch. Typically, it's advisable to promptly remove these potential lateral branches and not retain them. Below the first lateral branch on the main stem, each leaf bud has the potential to develop into a lateral branch. Typically, it's advisable to promptly remove these potential lateral branches and not retain them. When pruning eggplants, it's recommended to start pruning when the second lateral branch begins to grow. At this point, the leaf buds on the lower leaves have started to develop into lateral branches. As the leaves on the plant grow larger from the bottom to the top, the upper lateral branches tend to have more vigorous growth. When pruning eggplants, except for retaining the top two lateral branches, it's advisable to promptly remove the lower lateral branches on the main stem. Leave one leaf on each of these branches and remove all other leaves. This promotes healthier growth and better fruit development on the retained upper lateral branches. While pruning, it's also beneficial to remove some of the lower, older leaves to improve air circulation and reduce the risk of diseases. Typically, you should remove only a portion of the older, withered, or heavily affected leaves by pests and diseases. The method for removing leaves is as follows: When the eggplant fruits reach a diameter of 3–4 cm, remove the lower old leaves of the plants. When the fruits of Simendou eggplant reach a diameter of 3–4 cm, remove the lower old leaves of the plants. After this stage, it's generally not necessary to continue leaf removal. In situations where the growing season is short or space is limited, topping or pruning the eggplant plants can be advantageous. For large-fruited varieties, after the initial flowering of the eggplants, maintain 1–2 leaves above the fruits when topping, and remove all newly emerging lateral branches. Maintain 7 leaves per plant to concentrate nutrients and accelerate fruit development, aiming for early yields. For smaller-sized varieties, you can leave an additional branch after the initial flowering of the Simendou eggplants, meaning that after the initial flowering, there should be 4 branches remaining, with all other lateral branches removed to prevent overcrowding of leaves and branches.

(IV) Task Record Sheet

Task name		Guiding teacher	
Group No.		Group leader	
Time		Record content	
Signature of teacher		Date	

(V) Assessment Sheet

<table>
<tr><td>Task name</td><td colspan="4">Adjustment of melon vegetable plants</td><td>Group No.</td><td colspan="3"></td></tr>
<tr><td>Implementation date</td><td></td><td colspan="7">Process record for adjustment of melon vegetable plants, ________ pages in total</td></tr>
<tr><td>Evaluation item</td><td colspan="2">Evaluation content</td><td>Percentage</td><td>Teacher's mark</td><td>Student's mark</td><td>Score</td><td>Total</td></tr>
<tr><td rowspan="4">Process evaluation</td><td rowspan="4">Working attitude</td><td>Work attendance</td><td>2%</td><td>1%</td><td>1%</td><td></td><td></td></tr>
<tr><td>Responsible attitude</td><td>3%</td><td>2%</td><td>1%</td><td></td><td></td></tr>
<tr><td>Communication</td><td>2%</td><td>1%</td><td>1%</td><td></td><td></td></tr>
<tr><td>Teamwork</td><td>3%</td><td>2%</td><td>1%</td><td></td><td></td></tr>
</table>

continued

Evaluation item	Evaluation content		Percentage	Teacher's mark	Student's mark	Score	Total
Process evaluation	Working method	Learning ability	3%	1%	2%		
		Planning ability	3%	2%	1%		
		Problem-solving ability	4%	3%	1%		
	Practical operation	Completeness of the preparation work	2%	1.5%	0.5%		
		Rationality of pruning	13%	12%	1%		
		Timeliness of pruning work	5%	4%	1%		
		Fruit-setting nodes	5%	4%	1%		
		Pruning location	3%	2%	1%		
		Timing for leaf removal	2%	1.5%	0.5%		
		Overall proficiency in operations	10%	7%	3%		
Result evaluation	Plant adjustment results	Overall effectiveness	10%	8%	2%		
		Analyzing the rationality of plant adjustment methods	10%	8%	2%		
	Training report	Correctness and compliance of report filling	20%	16%	4%		

VIII. Task Implementation Phase 7: Vegetable Fertilization and Irrigation Management

(I) Purpose of the Implementation of the Task Phase

Based on the growth status and growth stage of vegetables, acquire an understanding of commonly used irrigation and drainage techniques as well as fertilization methods for vegetables.

(II) Materials and Tools

Vegetable plants, fertilizers, farming tools, etc.

(III) Methods and Steps

Each group should perform the following operations according to the requirements.

1. Fertilization and irrigation management of melon vegetables

Cucumbers require a significant amount of nitrogen and potassium, especially during the fruiting period. After the first cucumber harvest, it's important to provide additional topdressing. Apply 20 kg of compound fertilizer containing potassium sulfate per mu. During the peak of cucumber harvesting, you can create furrows between the beds and apply 15 kg of compound fertilizer containing potassium sulfate per mu. After applying the fertilizer, cover it with soil. In the anaphase, consider supplementing the soil with extraradical topdressing through root application. This can include the application of micronutrient fertilizers such as potassium dihydrogen phosphate, magnesium sulfate, boron, or a 1% urea solution to promote cucumber growth. For plants with weaker growth, increase the amount of fertilizer and the frequency of topdressing. Moisture management: Regarding water management, it's crucial to ensure an adequate water supply during the flowering and fruiting stages.

2. Fertilization and irrigation management of solanaceous vegetables

(1) Tomato: For fertilization, first, ensure an adequate basal fertilizer application. Before winter, apply 5,000 kg of organic manure per mu and plow it into the soil to a depth of 35–40 cm. Second, Apply phosphorus fertilizer during transplanting. For each mu, apply 30–40 kg of calcium superphosphate in the

planting furrow as a basal fertilizer. Mixing it with well-fermented organic manure can enhance its effectiveness. Third, after the first fruit sets, conduct the first round of topdressing. For each mu, apply 500 kg of well-rotted dry manure or 25–30 kg of ammonium bicarbonate in the furrows. Also, scatter 100 kg of wood ash, then hoe and water the soil. Fourth, after harvesting the first fruit cluster, proceed with the second round of topdressing. Apply 15–20 kg of ammonium bicarbonate per mu. In conditions where available, you can also apply some human urine. Fifth, after harvesting the second fruit cluster, conduct the third round of topdressing. Apply 15 kg of ammonium bicarbonate per mu. Additionally, in areas with severe virus diseases, it's advisable to apply a small amount of quick-acting nitrogen fertilizer after the slow seedling growth period following transplanting. When combined with watering, this can help reduce disease incidence. Tomatoes require a significant amount of water throughout their entire growth period. During the fruiting stage, the optimal soil moisture content is 75%–90% of field water-holding capacity. However, for spring-planted tomatoes after the slow seedling growth period, controlling watering to some extent and frequent hoeing can promote root system growth and fruit setting. After the first fruit sets and with rising temperatures, it's important to combine topdressing with watering. Tomatoes are not tolerant of waterlogging, so avoid water stagnation when irrigating and ensure proper drainage after rainfall. Topdressing can be combined with watering. Alternately spray a 0.5% urea solution and a 0.3% solution of potassium dihydrogen phosphate, or alternate between a 0.5% urea solution and a 1%–2% solution of calcium superphosphate extract, along with a 2%–5% solution of wood ash extract.

(2) Eggplant: Eggplants have well-developed root systems and vigorous growth. They have high requirements for fertilization and irrigation. They are considered nutrient-demanding plants, not drought-tolerant, and intolerant of waterlogging. Therefore, it's essential to strengthen fertilization and irrigation management for eggplants. Before fruit setting, apply nitrogen fertilizer sparingly, and increase the application after fruit set. Generally, apply additional fertilizer every 5–7 days. Alternatively, you can apply fertilizer after each fruit harvest.

For topdressing, apply 50 kg of compound fertilizer and 30 kg of potassium fertilizer per mu. Alternatively, apply composted human or livestock manure. You can also combine topdressing with soil preparation, applying 50 kg of compound fertilizer and 50 kg of urea per mu, along with 25 kg of composted peanut bran, and 50 kg of phosphorus and potassium fertilizers between the plants. To promote healthy growth, dark green foliage, improve fruit quality, extend harvesting period, and increase yields, foliar fertilization with products like potassium dihydrogen phosphate or green fenway can be used. During the peak fruiting stage, maintain adequate but not excessive soil moisture. Avoid fluctuating between excessively dry and waterlogged conditions. Eggplants have a long harvesting period, so it's essential to ensure a consistent water and nutrient supply even after the initial fruit harvest to prevent the development of small or misshapen fruits.

(3) Pepper: Peppers require slightly less water compared to tomatoes and eggplants. During transplanting, only provide enough water in the planting furrows. After watering, avoid immediately sealing the furrows, allowing for better contact and increased ground temperature. After 4–5 days, water the furrows along the planting rows once with slow seedling water and then level the furrows. Subsequently, perform inter-row cultivation and soil loosening 3–4 times between the plants to maintain moisture, raise soil temperature, and facilitate seedling growth. Until just before chili pepper harvesting, it's best to refrain from excessive watering. After harvesting chili peppers, commence with topdressing and irrigation, and engage in inter-row soil hilling. 10 kg of urea, 20 kg of calcium superphosphate, or 15 kg of compound fertilizer can be applied per mu. Potassium fertilizer has a significant effect on increasing chili pepper yields and disease resistance but should be applied in moderation. During the peak fruiting stage, perform topdressing 3–4 times as mentioned above and accompany each application with irrigation between the rows 1–2 times. After the crop closure stage, maintain the soil in a consistently moist state. During the rainy season, ensure proper drainage in case of heavy rainfall. Additionally, after the peak fruiting period, foliar spray with 0.2%–0.3% potassium

dihydrogen phosphate is recommended.

3. Fertilization and irrigation management of legumes

For legumes, it's essential to practice precise and efficient irrigation. Avoid heavy flooding. During the seedling stage, control the amount of water strictly. After applying thorough bottom-watering and slow seedling watering, reduce or limit further irrigation. Increase the volume of water significantly after the first flower set to promote pod growth. During the peak pod development period, gradually decrease the amount of water to prevent flower and pod drop. In low temperatures, use root-promoting fertilizers such as seaweed extract, chitosan, and humic acid to stimulate fine root growth. Reduce the application of compound fertilizers as much as possible and consider using fully water-soluble fertilizers as a replacement. After the initial basal fertilization, additional fertilization is not required until after the first flower set. At that point, fertilization can be applied as needed. However, once plants enter the peak pod development stage, it's advisable to limit further heavy fertilization. Instead, apply smaller amounts of fertilizer multiple times in the anaphase is more beneficial for root absorption.

4. Fertilization and irrigation management of cabbage vegetables

Cabbage and similar vegetables have relatively high requirements for fertilization and irrigation. For small cabbage, start applying slow seedling fertilizer 3–5 days after transplanting to promote rapid growth. Fertilize every 5–7 days, and stop topdressing 15–20 days before harvesting. Typically, apply 10–20 kg of urea per mu. For Chinese cabbage, the main fertilization occurs during the head formation stage. Apply well-rotted manure at a rate of 1,000 kg per mu, ammonium dihydrogen phosphate at 2 kg, and potassium sulfate at 20 kg. Common head cabbage has a strong tolerance to fertilization, so in addition to sufficient basal fertilization, it's important to continue with topdressing. For cauliflower during head formation, apply compound fertilizer 1–2 times, typically at a rate of 20 kg per mu.

5. Fertilization and irrigation management of root vegetables

For root vegetables, it is essential to apply one-time basal fertilization before plowing. Apply 300 kg of well-rotted organic fertilizer and 3 kg of compound fertilizer per 180 m^2 of land. It is crucial to emphasize that the organic fertilizer used must be thoroughly decomposed and fermented. Basal fertilization should be combined with plowing the soil 7–10 days before sowing. The fertilization principle prioritizes basal fertilization with supplemental topdressing. Basal fertilization promotes root growth, topdressing encourages leaf development, and basal fertilization promotes head formation. It is recommended to water 2–3 times in the early stage to stimulate the growth of rosette leaves, control watering appropriately during the middle and late stages, and perform frequent cultivation to loosen the soil to prevent excessive leaf elongation. After the fleshy roots are exposed, it is crucial to perform a significant topdressing combined with intertillage and irrigation. About 100 kg of wood ash and 30–40 kg of microbial compound fertilizer should be applied between the rows per mu. During this period, ensure uniform water and fertilizer supply, typically watering once every 5–7 days and applying compound fertilizer every 2 weeks.

6. Fertilization and irrigation management of leafy vegetables

(1) Lettuce and bamboo shoots.

Overwintering with seedlings, the early growth is slow, and nutrient requirements are low. Before winter, it's necessary to control fertilization and irrigation to prevent excessive elongation and enhance cold resistance. After the onset of spring, the stems and leaves grow rapidly. When entering the heart-forming stage, apply a single topdressing. Apply 1,000 kg of 30% human manure per mu. After the crop closure stage, when the stems need more nutrients for bulking, apply topdressing 2–3 times. In total, apply 2,000–3,000 kg of human manure or 30–40 kg of urea to ensure the stems swell. Fertilization should not be delayed to prevent stem cracking.

(2) Celery.

When the inner leaves of celery begin vigorous growth, apply quick-acting nitrogen fertilizer, around 10 kg of ammonium sulfate per mu. Avoid applying quick-acting nitrogen fertilizer 30 days before harvest. If there is a water shortage, ensure proper irrigation.

7. Fertilization and irrigation management of allium vegetables

(1) Onion fertilization and irrigation management.

For the first time, it is recommended to apply topdressing fertilizer during the regreening phase (late February or early March). Combine it with regrowth water. Apply approximately 1,000–1,500 kg of composted manure or 10 kg of urea per mu. Additionally, 20 kg of calcium dihydrogen phosphate should be added to stimulate regrowth and sprouting. Around 30 days after regreening (around mid-April), when the foliage enters a vigorous growth stage, there is a greater need for fertilizer, particularly nitrogen (N). Apply a heavy topdressing of approximately 2,500 kg of composted manure or 15 kg of urea per mu. Complement this with 2–3 kg of potassium dihydrogen phosphate or 10 kg of potassium sulfate. Approximately 50–60 days after regreening (around mid-May), when the onion bulbs start to enlarge, it's a critical period for topdressing. Apply around 10–25 kg of ammonium sulfate or about 20 kg of urea per mu. Add 3–4 kg of potassium dihydrogen phosphate or approximately 15 kg of potassium sulfate. This aids in the development of bulbs and enhances their quality and storage ability.

(2) Garlic fertilization and irrigation management.

Topdressing for garlic is carried out in two stages. The first application is conducted around the time of maternal plant withering, and it should be combined with irrigation. Apply approximately 500 kg of composted manure or 20 kg of potassium sulfate compound fertilizer per mu. The second application should take place after the garlic plants have started to grow scapes, again combined with irrigation. Apply approximately 10 kg of urea or 15 kg of potassium sulfate compound fertilizer per mu.

(3) Chinese chive fertilization and irrigation management.

Before late summer (around the beginning of August), it is generally not advisable to carry out additional topdressing and watering. In the middle of August, as temperatures begin to drop, Chinese chives enter a period of vigorous growth. During this time, it is recommended to apply additional fertilizer. 200–250 kg of well-decomposed soybean cake fertilizer can be applied per mu.

8. Fertilization and irrigation management of tuberous vegetables

(1) Potato fertilization and irrigation management.

After the seedlings emerge, apply strong seedling fertilizer at a rate of 105–120 kg of urea and potassium sulfate per hm^2. When the potato seedlings reach a height of 15–20 cm, apply rooting fertilizer at a rate of 150 kg of potassium sulfate and 105–120 kg of urea per 10,000 m^2. In the anaphase, adjust the fertilization based on the colors of the plant. When the plant leaves are dark green, 225 kg of potassium fertilizer can be applied per 10,000 m^2. If the plants are robust and the stems are golden-yellow in color, additional fertilization is not required.

(2) Taro fertilization and irrigation management.

With a sufficient base fertilizer, diluted manure water can be applied in the early stages. When the taro begins to swell, perform one round of topdressing. 1,000 kg of farmyard manure, 50 kg of peanut shells, 15 kg of potassium sulfate, 50 kg of organic biological fertilizer, and 2 kg of boron-zinc-magnesium fertilizer can be mixed together and applied per mu.

(3) Ginger fertilization and irrigation management.

Regarding fertilization, when ginger seedlings are around 30 cm tall, 10 kg of urea should be applied per mu. Around the time of the autumn equinox, in combination with soil hilling, apply 10 kg of urea and 10 kg of compound fertilizer per mu. In early September, when ginger plants have 6–8 branches, if the soil fertility is poor and plant growth is average, apply 20 kg of compound fertilizer along with 5 kg of urea per mu. As for watering, it's important to thoroughly water the soil before planting. During the seedling stage, apply minimal water. During the vigorous growth period after the autumn equinox, water generously, keeping the soil consistently moist. Additionally, during the rainy season, ensure that any water accumulation in the furrows and ditches is promptly drained.

(IV) Task Record Sheet

Task name		Guiding teacher	
Group No.		Group leader	
Time		Record content	
Signature of teacher		Date	

(V) Assessment Sheet

<table>
<tr><td>Task name</td><td colspan="4">Vegetable fertilization and irrigation management</td><td>Group No.</td><td colspan="3"></td></tr>
<tr><td>Implementation date</td><td colspan="2"></td><td colspan="6">Process record for vegetable fertilization and irrigation management, ________ pages in total</td></tr>
<tr><td>Evaluation item</td><td colspan="2">Evaluation content</td><td>Percentage</td><td>Teacher's mark</td><td>Student's mark</td><td>Score</td><td>Total</td></tr>
<tr><td rowspan="4">Process evaluation</td><td rowspan="4">Working attitude</td><td>Work attendance</td><td>2%</td><td>1%</td><td>1%</td><td></td><td></td></tr>
<tr><td>Responsible attitude</td><td>3%</td><td>2%</td><td>1%</td><td></td><td></td></tr>
<tr><td>Communication</td><td>2%</td><td>1%</td><td>1%</td><td></td><td></td></tr>
<tr><td>Teamwork</td><td>3%</td><td>2%</td><td>1%</td><td></td><td></td></tr>
</table>

continued

Evaluation item	Evaluation content		Percentage	Teacher's mark	Student's mark	Score	Total
Process evaluation	Working method	Learning ability	3%	1%	2%		
		Planning ability	3%	2%	1%		
		Problem-solving ability	4%	3%	1%		
	Practical operation	Completeness of the preparation work	2%	1.5%	0.5%		
		Rationality of fertilization program development	6%	3%	3%		
		Calculation of organic fertilizer application quantity	10%	8%	2%		
		Calculation of chemical fertilizer usage	5%	4%	1%		
		Fertilization timing	10%	8%	2%		
		Fertilization methods and procedures	7%	5%	2%		
Result evaluation	Fertilization result	Overall effectiveness	10%	8%	2%		
		Analyzing the rationality of plant fertilization	10%	8%	2%		
	Training report	Correctness and compliance of report filling	20%	16%	4%		

IX. Task Implementation Phase 8: Hand Pollination and Fruit Retention in Melon Vegetables

(I) Purpose of the Implementation of the Task Phase

Learn the techniques of hand pollination and fruit retention in melon vegetables according to their growth and development stages.

(II) Materials and Tools

Melon vegetables, labels, paintbrushes, farming tools, and Baoguoling.

(III) Methods and Steps

Each group should perform the following operations according to the requirements.

1. Hand pollination and fruit retention in watermelons

(1) Select the right position for fruit retention. The choice of fruit retention positions depends on the varieties and weather conditions. For small-sized watermelons, when the main vine has 18–20 nodes, select the second pistillate flower at 10th–12th nodes or the first pistillate flower on a lateral branch for pollination. For medium-sized watermelons, when the main vine has 20–25 nodes, choose the second pistillate flower at 13th–15th nodes or the first pistillate flower on a lateral branch for pollination. In the case that plant growth vigor is weak, fruit retention can be slightly delayed, and conversely, fruit retention positions can be advanced.

(2) Selecting pistillate and staminate flowers. The quality of pistillate flowers has a significant impact on fruit development. Pistillate flowers should have long and stout pedicels, well-developed ovaries, a normal appearance, tender green skin with a glossy sheen, dense trichomes, and good development. Staminate flowers should be healthy, free from diseases, fully mature, and have abundant pollen. It is best to choose staminate and pistillate flowers that have opened on the same day for hand pollination, as this results in the highest fertilization rate.

(3) Determine the pollination timing. The optimal time for pollination of greenhouse watermelons is 7:00–9:00 on a sunny day, as this is when the styles of the pistillate flowers and the physiological activity of the staminate flower pollen are most vigorous. On rainy days, when staminate flowers release pollen later, you

can consider delaying pollination or collecting staminate flowers that will open the next day in the previous afternoon. Keep them indoors in dry and warm conditions, and pollinate the pistillate flowers the next morning after they have opened.

(4) Hand pollination methods. ① Flower-to-flower method: Take the open staminate flowers, remove the petals or fold them back with tweezers to expose the stamen, then gently lift the pistillate flowers, exposing the stigma. Rub the stamen onto the stigma lightly to transfer the pollen. Pollen application should be even to avoid the development of malformed fruits. Generally, one staminate flower can be used to pollinate 2–3 pistillate flowers. ② Brush dabbing method: Collect pollen from freshly opened staminate flowers into a clean container. Then, use a soft brush to dab the pollen and apply it gently to the stigma of pistillate flowers. Visible yellow pollen on the stigma indicates successful pollination. After pollination, label the date or use differently colored sticks to mark the pollinated flowers for batch harvesting of ripe fruits.

(5) Checking pollination effect. The day after pollination, in the afternoon, if the fruit stalk of the pistillate flower has elongated or bent, and the ovary has significantly swollen, it indicates successful pollination. If the fruit stalk remains upright or facing forward, it means that pollination was unsuccessful. In this case, you should perform another round of hand pollination on the third pistillate flower of the main vine or the second pistillate flower of a lateral vine.

(6) During the peak flowering period of greenhouse watermelons, it's essential to maintain sufficient light and relatively high night temperatures. After hand pollination, if the night temperatures are low, it can lead to fruit drop or hinder fruit enlargement. Approximately 7 days after pollination, when the fruit has set, ample nutrients and water should be applied. When the young fruit reaches the size of an egg, select one well-developed, well-shaped young fruit per plant (usually around the 12th to 14th leaf on the main vine) and gently twist the main vine three nodes above the selected fruit to control excessive growth and promote fruit enlargement.

(7) In recent years, in some regions, the use of bees for pollination in spring greenhouses for small watermelons has replaced traditional methods involving manual pollination assistance and the use of growth regulators to promote the fruit set. This approach has effectively addressed the challenges of poor pollination

and low fruit set in greenhouse-grown watermelons. It not only saves labor but also improves the quality of watermelons. The resulting fruits are round and well-shaped, with good surface smoothness, crisp and juicy flesh, and high sugar content.

2. Hand pollination and fruit retention in netted melons

(1) Plant regulation. After the seedlings have turned their vines, use plastic twine to support the vines, single-stem pruning, and fruit setting on lateral vines. Begin fruit thinning at the 12th–14th node, leaving 2–3 fruit branches per plant. Remove all other lateral branches. On lateral vines, leave one fruit and two leaves before pruning the growing point. On the main vine, top it at around the 26th node, and promptly remove newly growing lateral branches.

(2) Pollination and fruit retention. Around 18 days after transplanting, when the 12th node of the lateral vine has pistillate flowers in bloom, perform hand pollination in the morning, pollinating each plant with 3–5 fruits. Once the fruits are firmly set, select and keep 1–2 young fruits with proper shape, small navel, long stems, and elliptical shape. When the fruits reach the size of a fist, it's important to thin them out promptly.

(3) Bagging the fruits. After determining which fruits to keep, bag them using colorless, transparent polyethylene bags that are approximately 600 mm long, 220 mm wide, and 0.08 mm thick.

(IV) Task Record Sheet

Task name		Guiding teacher	
Group No.		Group leader	
Time		Record content	
Signature of teacher		Date	

(V) Assessment Sheet

<table>
<tr><td>Task name</td><td colspan="4">Flower and fruit retention of melon and solanaceous vegetable plants</td><td>Group No.</td><td colspan="2"></td></tr>
<tr><td>Implementation date</td><td></td><td colspan="6">Process record for flower and fruit retention of melon and solanaceous vegetable plants, ________ pages in total</td></tr>
<tr><td>Evaluation item</td><td colspan="2">Evaluation content</td><td>Percentage</td><td>Teacher's mark</td><td>Student's mark</td><td>Score</td><td>Total</td></tr>
<tr><td rowspan="17">Process evaluation</td><td rowspan="4">Working attitude</td><td>Work attendance</td><td>2%</td><td>1%</td><td>1%</td><td></td><td></td></tr>
<tr><td>Responsible attitude</td><td>3%</td><td>2%</td><td>1%</td><td></td><td></td></tr>
<tr><td>Communication</td><td>2%</td><td>1%</td><td>1%</td><td></td><td></td></tr>
<tr><td>Teamwork</td><td>3%</td><td>2%</td><td>1%</td><td></td><td></td></tr>
<tr><td rowspan="3">Working method</td><td>Learning ability</td><td>3%</td><td>1%</td><td>2%</td><td></td><td></td></tr>
<tr><td>Planning ability</td><td>3%</td><td>2%</td><td>1%</td><td></td><td></td></tr>
<tr><td>Problem-solving ability</td><td>4%</td><td>3%</td><td>1%</td><td></td><td></td></tr>
<tr><td rowspan="6">Practical operation</td><td>Rationality of hormone preparation concentration</td><td>10%</td><td>8%</td><td>2%</td><td></td><td></td></tr>
<tr><td>Rationality of configuring and using equipment</td><td>5%</td><td>4%</td><td>1%</td><td></td><td></td></tr>
<tr><td>Proficiency of the configuration process</td><td>5%</td><td>4%</td><td>1%</td><td></td><td></td></tr>
<tr><td>Timing of hormone application</td><td>10%</td><td>6%</td><td>4%</td><td></td><td></td></tr>
<tr><td>Sense of labor protection</td><td>2%</td><td>1.5%</td><td>0.5%</td><td></td><td></td></tr>
<tr><td>Correctness of hormone usage procedures</td><td>8%</td><td>6%</td><td>2%</td><td></td><td></td></tr>
</table>

continued

Evaluation item	Evaluation content		Percentage	Teacher's mark	Student's mark	Score	Total
Result evaluation	Result of flower and fruit retention	Overall effectiveness	10%	8%	2%		
		Absence of flower and fruit drop	10%	8%	2%		
	Training report	Correctness and compliance of report filling	20%	16%	4%		

X. Task Implementation Phase 9: Flower and Fruit Retention of Melon and Solanaceous Vegetable Plants

(I) Purpose of the Implementation of the Task Phase

Understand the reasons for flower and fruit drop in solanaceous vegetables and master the application methods of plant growth regulators in flower and fruit retention of solanaceous vegetables.

(II) Materials and Tools

Solanaceous flowering vegetable plants, 2,4-D, PCPA, small sprayer, paintbrush, dye, etc.

(III) Methods and Steps

Each group should perform the following operations according to the requirements.

1. Tomato flower and fruit retention techniques

(1) Preparation of chemical solutions.

Prepare 50 mL of solution per person. Mix 2,4-D to create a solution with a concentration of 15–25 mg/L, and PCPA to make a solution with a concentration of 10–50 mg/L.

(2) Application methods.

① 2,4-D can be applied to the flowers from the initial flowering stage to the full flowering stage. Dip the selected flowers in the chemical solution for a brief moment. Each flower should be dipped once, or you can use a brush to apply the solution at the base of the pedicel abscission layer.

② PCPA should be applied when there are 3–4 or more flowers open within one inflorescence of tomatoes. It should be sprayed on all the open flowers and unopened flower buds on the same inflorescence.

2. Eggplant flower and fruit retention techniques

Eggplants experience flower and fruit drop to varying degrees during their growth. The reasons for this include not only high and low temperatures (above 38°C or below 15°C) but also factors like insufficient light, dry soil, poor nutrition, and defects in the flower structure. To prevent flower and fruit drop in eggplants, it's important to enhance field management based on the underlying causes and improve the nutritional status of the plants. Additionally, plant growth regulators can effectively prevent temperature-induced flower and fruit drop.

2,4-D can be used to dip or paint the flowers. To do this, prepare a 30–40 g/L dilution of 2,4-D and place it in a small bowl. Then, dip the flowers into the solution, ensuring they reach the base of the pedicel, and immediately remove them. Alternatively, you can use a fine brush to apply the solution to the flowers, taking care to remove any excess liquid from the stem to avoid causing deformities due to high concentration. Prepare and use the solution immediately, preferably on clear, sunny days. Adjust the concentration depending on the temperatures (Use a slightly higher concentration in cooler weather and a slightly lower concentration in warmer weather). Typically, apply the treatment at 8:00–10:00 in the morning or 14:00–16:00 in the afternoon. A 30–50 g/L solution of Fanqieling or a Fangluosu can also be used for spraying on the flowers, ensuring even coverage.

3. Pepper flower and fruit retention techniques

During periods of low temperatures when chili pepper plants are prone to flowering and fruit drop, chemical agents can also be used to protect the flowers and fruits. During the flowering stage, a 20–30 parts per million (PPM) solution of 2,4-D or a 25–30 PPM solution of Fanqieling can also be used for spraying on the flowers or for applying to the pedicel by brushing. It's recommended to carry out the spraying at 8:00–11:00 in the morning.

(IV) Task Record Sheet

<table>
<tr><td>Task name</td><td></td><td>Guiding teacher</td><td></td></tr>
<tr><td>Group No.</td><td></td><td>Group leader</td><td></td></tr>
<tr><td>Time</td><td></td><td>Record content</td><td></td></tr>
<tr><td></td><td colspan="3"></td></tr>
<tr><td></td><td colspan="3"></td></tr>
<tr><td></td><td colspan="3"></td></tr>
<tr><td></td><td colspan="3"></td></tr>
<tr><td></td><td colspan="3"></td></tr>
<tr><td></td><td colspan="3"></td></tr>
<tr><td></td><td colspan="3"></td></tr>
<tr><td>Signature of teacher</td><td></td><td>Date</td><td></td></tr>
</table>

(V) Assessment Sheet

<table>
<tr><td>Task name</td><td colspan="4">Flower and fruit retention of melon and solanaceous vegetable plants</td><td>Group No.</td><td colspan="3"></td></tr>
<tr><td>Implementation date</td><td></td><td colspan="7">Process record for flower and fruit retention of melon and solanaceous vegetable plants, ________ pages in total</td></tr>
<tr><td>Evaluation item</td><td colspan="2">Evaluation content</td><td>Percentage</td><td>Teacher's mark</td><td>Student's mark</td><td>Score</td><td>Total</td></tr>
<tr><td rowspan="4">Process evaluation</td><td rowspan="4">Working attitude</td><td>Work attendance</td><td>2%</td><td>1%</td><td>1%</td><td></td><td></td></tr>
<tr><td>Responsible attitude</td><td>3%</td><td>2%</td><td>1%</td><td></td><td></td></tr>
<tr><td>Communication</td><td>2%</td><td>1%</td><td>1%</td><td></td><td></td></tr>
<tr><td>Teamwork</td><td>3%</td><td>2%</td><td>1%</td><td></td><td></td></tr>
</table>

continued

Evaluation item	Evaluation content		Percentage	Teacher's mark	Student's mark	Score	Total
Process evaluation	Working method	Learning ability	3%	1%	2%		
		Planning ability	3%	2%	1%		
		Problem-solving ability	4%	3%	1%		
	Practical operation	Rationality of hormone preparation concentration	10%	8%	2%		
		Rationality of configuring and using equipment	5%	4%	1%		
		Proficiency of the configuration process	5%	4%	1%		
		Timing of hormone application	10%	6%	4%		
		Sense of labor protection	2%	1.5%	0.5%		
		Correctness of hormone usage procedures	8%	6%	2%		
Result evaluation	Result of flower and fruit retention	Overall effectiveness	10%	8%	2%		
		Absence of flower and fruit drop	10%	8%	2%		
	Training report	Correctness and compliance of report filling	20%	16%	4%		

XI. Task Implementation Phase 10: Vegetable Pest and Disease Control

(I) Purpose of the Implementation of the Task Phase

Understanding the reasons for the occurrence of major diseases and pests in vegetables and mastering various comprehensive methods for pest and disease control is crucial.

(II) Materials and Tools

Vegetable plants, agricultural sprayers, various pesticides, masks, latex gloves, measuring cups, etc.

(III) Methods and Steps

Each group should perform the following operations according to the requirements.

1. Basic information investigation

(1) Understand and familiarize yourself with common vegetable diseases and pests.

(2) Understand the infection pathways and development patterns of major vegetable diseases.

(3) Understand and grasp the types, occurrences, and patterns of major vegetable diseases.

(4) Understand the impact of local climatic conditions on the growth and development patterns of vegetables and the occurrence and development of pests and diseases.

(5) Know the types and usage of common pesticides in the local area.

2. Establishing principles and requirements

(1) Implement the principle of "prevention first, comprehensive control," comprehensively use various control measures, effectively manage biological hazards, and keep pesticide residues within the prescribed standards.

(2) In consideration of the increasing awareness of green consumption among the domestic population, the exceptionally strict international trade pesticide residue detection standards, and the protection of technological green barriers, it is proposed to improve traditional methods in the production of melons and move towards

precision approaches.

(3) Starting from local conditions, with clear objectives, specific content, and practical feasibility.

(IV) Task Record Sheet

<table>
<tr><td>Task name</td><td></td><td>Guiding teacher</td><td></td></tr>
<tr><td>Group No.</td><td></td><td>Group leader</td><td></td></tr>
<tr><td>Time</td><td></td><td>Record content</td><td></td></tr>
<tr><td></td><td colspan="3"></td></tr>
<tr><td></td><td colspan="3"></td></tr>
<tr><td></td><td colspan="3"></td></tr>
<tr><td></td><td colspan="3"></td></tr>
<tr><td></td><td colspan="3"></td></tr>
<tr><td></td><td colspan="3"></td></tr>
<tr><td>Signature of teacher</td><td></td><td>Date</td><td></td></tr>
</table>

(V) Assessment Sheet

<table>
<tr><td>Task name</td><td colspan="5">Vegetable pest and disease control</td><td colspan="2">Group No.</td><td></td></tr>
<tr><td>Implementation date</td><td></td><td colspan="7">Process record for vegetable pest and disease control, ________ pages in total</td></tr>
<tr><td>Evaluation item</td><td colspan="2">Evaluation content</td><td>Percentage</td><td>Teacher's mark</td><td>Student's mark</td><td>Score</td><td>Total</td></tr>
<tr><td rowspan="4">Process evaluation</td><td rowspan="4">Working attitude</td><td>Work attendance</td><td>2%</td><td>1%</td><td>1%</td><td></td><td></td></tr>
<tr><td>Responsible attitude</td><td>3%</td><td>2%</td><td>1%</td><td></td><td></td></tr>
<tr><td>Communication</td><td>2%</td><td>1%</td><td>1%</td><td></td><td></td></tr>
<tr><td>Teamwork</td><td>3%</td><td>2%</td><td>1%</td><td></td><td></td></tr>
</table>

continued

Evaluation item	Evaluation content		Percentage	Teacher's mark	Student's mark	Score	Total
Process evaluation	Working method	Learning ability	3%	1%	2%		
		Planning ability	3%	2%	1%		
		Problems-solving ability	4%	3%	1%		
	Practical operation	Accurate observation of pests and diseases with serious attitude	13%	8%	5%		
		Appropriate selection of medicines for specific pests and diseases, and accurate concentration of prepared pesticide solutions	15%	10%	5%		
		Accurate spraying timing, and even pesticide application, with attention to personal safety precautions	12%	6%	6%		
Result evaluation	Pest and disease control result	Overall effectiveness	10%	8%	2%		
		Pesticide residue	10%	8%	2%		
	Training report	Correctness and compliance of report filling	20%	16%	4%		

XII. Task Implementation Phase 11: Vegetable Harvesting and Post-harvest Handling

(I) Purpose of the Implementation of the Task Phase

Understand the main harvesting and post-harvest handling of vegetables, and master various integrated methods for pest and disease control.

(II) Materials and Tools

Vegetable fruits, containers, cold storage, waxing machine, packaging machine, etc.

(III) Methods and Steps

Each group should perform the following operations according to the requirements.

1. Harvesting

Melons, solanaceous vegetables, legumes, cabbage vegetables, root vegetables, allium vegetables, and tuberous vegetables must be harvested at the appropriate stage of ripeness to ensure the quality for packaging and storage. Most of the harvesting is done manually. Melons, solanaceous vegetables, legumes, cabbage vegetables, and leafy vegetables are harvested and packed into large crates, and then transported to the factory. Careful selection is made, and damaged or low-quality produce is removed. Harvesting is typically done in the early morning when temperatures are at their lowest. During this time, the harvested, sorted, and graded vegetables have lower temperatures, which can reduce the time and cost of pre-cooling and help maintain better quality.

2. Grading by size

The packing and temporary storage of melons, solanaceous vegetables, allium vegetables, and tuberous vegetables are gradually being shifted from packaging plants to on-farm packaging and transportation. Legumes are graded by the sizes of the fruits. Cabbage vegetables are graded by the sizes of the heads. Root vegetables are graded by lengths. Leafy vegetables are classified into three categories based on the commercial characteristics of different varieties.

3. Packaging, processing, or export

Packaging is not suitable for the strict grading of vegetables. Vegetables purchased by packaging factories are initially stored in a cool place, and any defective products should be immediately removed. They are graded, waxed, packaged in commercial packaging, and placed in a cold storage for cooling and storage until they are ready for the market or export. Packaging of rigorously graded leafy vegetables is done according to customer requirements.

(IV) Task Record Sheet

<table>
<tr><td>Task name</td><td></td><td>Guiding teacher</td><td></td></tr>
<tr><td>Group No.</td><td></td><td>Group leader</td><td></td></tr>
<tr><td>Time</td><td></td><td>Record content</td><td></td></tr>
<tr><td></td><td colspan="3"></td></tr>
<tr><td></td><td colspan="3"></td></tr>
<tr><td></td><td colspan="3"></td></tr>
<tr><td></td><td colspan="3"></td></tr>
<tr><td></td><td colspan="3"></td></tr>
<tr><td></td><td colspan="3"></td></tr>
<tr><td></td><td colspan="3"></td></tr>
<tr><td>Signature of teacher</td><td></td><td>Date</td><td></td></tr>
</table>

(V) Assessment Sheet

<table>
<tr><td>Task name</td><td colspan="5">Vegetable harvesting and post-harvest handling</td><td>Group No.</td><td colspan="2"></td></tr>
<tr><td>Implementation date</td><td></td><td colspan="7">Process record for vegetable harvesting and post-harvest handling, ______ pages in total</td></tr>
<tr><td>Evaluation item</td><td colspan="2">Evaluation content</td><td>Percentage</td><td>Teacher's mark</td><td>Student's mark</td><td>Score</td><td>Total</td></tr>
<tr><td rowspan="7">Process evaluation</td><td rowspan="4">Working attitude</td><td>Work attendance</td><td>2%</td><td>1%</td><td>1%</td><td></td><td></td></tr>
<tr><td>Responsible attitude</td><td>3%</td><td>2%</td><td>1%</td><td></td><td></td></tr>
<tr><td>Communication</td><td>2%</td><td>1%</td><td>1%</td><td></td><td></td></tr>
<tr><td>Teamwork</td><td>3%</td><td>2%</td><td>1%</td><td></td><td></td></tr>
<tr><td rowspan="3">Working method</td><td>Learning ability</td><td>3%</td><td>1%</td><td>2%</td><td></td><td></td></tr>
<tr><td>Planning ability</td><td>3%</td><td>2%</td><td>1%</td><td></td><td></td></tr>
<tr><td>Problems-solving ability</td><td>4%</td><td>3%</td><td>1%</td><td></td><td></td></tr>
</table>

continued

Evaluation item	Evaluation content		Percentage	Teacher's mark	Student's mark	Score	Total
Process evaluation	Practical operation	Determining the degree of ripeness	8%	7%	1%		
		Proficiency in harvesting operations	12%	10%	2%		
		Proficiency in post-harvest handling	12%	6%	6%		
		Proficiency in pre-cooling operations	8%	6%	2%		
Result evaluation	Harvesting and post-harvest handling results	Harvesting and post-harvest handling quantity	10%	5%	5%		
		Overall effectiveness	10%	6%	4%		
	Training report	Correctness and compliance of report filling	20%	16%	4%		

References

[1] China Agricultural Academy of Sciences, Institute of Vegetables and Flowers. Vegetable Cultivation in China [M]. Beijing: China Agricultural Press, 2010.

[2] YU X H. Vegetable Production Techniques and Practical Training [M]. Beijing: China Labor and Social Security Publishing House, 2005.

[3] LV J L. Vegetable Cultivation (Southern Edition) [M]. 3rd edition. Beijing: China Agricultural Press, 2001.

[4] HU F R. Vegetable Cultivation [M]. Shanghai: Shanghai Jiao Tong University Press, 2003.

[5] CHEN X Y. Vegetable Cultivation [M]. Beijing: Higher Education Press, 2010.

[6] HAN S D. Vegetable Production Techniques [M]. Beijing: China Agricultural Press, 2006.

[7] Zhejiang Agricultural Technology Promotion Center. Standardized Production Techniques for Watermelons and Cantaloupes [M]. Hangzhou: Zhejiang Science and Technology Press, 2008.

[8] JU J F. Practical Training and Assessment of Horticultural Skills [M]. Beijing: China Agricultural Press, 2006.

[9] DONG H X, GUI D P. Green Vegetable Production Techniques [M]. Beijing: China Agricultural University Press, 2010.

[10] LI Z X. Horticultural Practice [M]. Beijing: China Agricultural Press, 2004.

[11] Editorial Committee of China Agricultural Encyclopedia (Vegetable Volume). China Agricultural Encyclopedia (Vegetable Volume) [M]. Beijing:

China Agricultural Press, 1990.

[12] YU F M, CHEN D M. Vegetable Cultivation Techniques Manual [M]. Shanghai: Shanghai Science and Technology Press, 2010.

[13] LI C G, BAO Y Z. New Issues and Countermeasures for the Development of China's Vegetable Industry [J]. China Vegetables, 2010, (15).

[14] Zhejiang Agricultural Technology Promotion Center. Standardized Vegetable Production Techniques [M]. Hangzhou: Zhejiang Science and Technology Press, 2008.

[15] HAN Z H, CHEN K S. Experimental Horticulture [M]. Beijing: Higher Education Press, 2006.

[16] HAN Q P, WANG B H. Vegetable Pests and Diseases Diagnosis and Control Techniques [M]. Beijing: Jindun Press, 2009.

[17] MAO M H. Practical Manual for Vegetable Production [M]. Shanghai: Shanghai Popular Science Press, 2008.

[18] WANG F H, CHEN S C. Standardized Vegetable Production Techniques [M]. Shanghai: Shanghai Science and Technology Press, 2007.

[19] LI Z L. Comprehensive Practical Training Course in Plant Production [M]. Beijing: China Agricultural Press, 2003.